Energy Auditing in Electrical Utilities

Energy Auditing in Electrical Utilities

Rajiv Shankar

Viva Books
New Delhi | Mumbai | Chennai | Kolkata
Bangalore | Hyderabad | Kochi | Guwahati

Publisher's note

Every possible effort has been made to ensure that the information contained in this book is accurate at the time of going to press, and the publisher and author cannot accept responsibility for any errors or omissions, however caused. No responsibility for loss or damage occasioned to any person acting, or refraining from action, as a result of the material in this publication can be accepted by the editor, the publisher or the author.

Every effort has been made to trace the owners of copyright material used in this book. The author and the publisher will be grateful for any omission brought to their notice for acknowledgement in the future editions of the book.

First Published 2010
Reprinted 2015, 2017, 2018, 2024

Viva Books Private Limited

- 4737/23, Ansari Road, Daryaganj, New Delhi 110 002
  Tel. 42242200, 23258325, 23283121, Email: vivadelhi@vivagroupindia.net
- 76, Service Industries, Shirvane, Sector 1, Nerul, Navi Mumbai 400 706
  Tel. 27721273, 27721274, Email: vivamumbai@vivagroupindia.net
- Megh Towers, Old No. 307, New No. 165, Poonamallee High Road, Maduravoyal
  Chennai 600 095 Tel. 23780991, 23780992, Email: vivachennai@vivagroupindia.net
- B-103, Jindal Towers, 21/1A/3 Darga Road, Kolkata 700 017
  Tel. 22816713, Email: vivakolkata@vivagroupindia.net
- 7, GF, Sovereign Park Aptts., 56-58, K. R. Road, Basavanagudi, Bengaluru 560 004
  Tel. 26607409, Email: vivabangalore@vivagroupindia.net
- 101-102, Moghal Marc Apartments, 3-4-637 to 641, Narayanguda, Hyderabad 500 029
  Tel. 27564481, Email: vivahyderabad@vivagroupindia.net
- First Floor, Beevi Towers, SRM Road, Kaloor, Kochi 682 018
  Tel. 0484-2403055, 2403056, Email: vivakochi@vivagroupindia.net
- 232, GNB Road, Beside UCO Bank, Silpukhuri, Guwahati 781 003
  Tel. 0361-2666386, Email: vivaguwahati@vivagroupindia.net

www.vivagroupindia.com

ISBN: 978-81-309-1469-5

Published by Vinod Vasishtha for Viva Books Private Limited, 4737/23 Ansari Road, Daryaganj, New Delhi 110 002.

Printed and bound in India by Replika Press Pvt. Ltd.

Contents

Acknowledgements

I wish to thank all my colleagues, friends and family members for their support in writing this book.

I am also thankful to the publisher Viva Books Private Limited for publishing the book.

Introduction

This book covers all the aspects of energy auditing in electrical utilities. Software is being supplied along with the book which can perform all calculations related with the energy audit. The software is user friendly and menu driven. By using the software various energy saving options can be analysed. The software has capability of generating results of energy saving analysis in a report form, which can be printed and attached with energy audit report.

The basic requirement of the software is to prepare a database of various parameters of equipments, which are normally collected during energy auditing phase. Besides this, data of energy efficient equipments like motors, pumps etc. are required to be stored, so that a comparison of operating cost of existing equipments and the energy efficient equipments can be carried out. The stored data is retrieved based on query generation. Retrieved data can be transferred from one screen to another screen for analysis purpose. Various other calculations can be performed just by clicking mouse of computer. A step by step procedure of performing above mentioned tasks is elaborated at the end of each chapter.

This book is useful for energy auditors, plant engineers, contractors, students and all those who are concerned with performance monitoring and improvement of plant equipments. A brief description of each chapter is given below.

Chapter 1: General Aspects of Energy Auditing

Chapter 1 highlights the benefits, which can be derived by performing energy auditing. The requirements and methodology of energy auditing, preparation of tender documents for awarding work of energy auditing to external energy auditor are discussed. This chapter also highlights the information, which is required to be incorporated in the energy audit report.

Chapter 2: Squirrel Cage Induction Motor

Chapter 2 deals with various aspects of induction motor like working principle, construction, affect of various external factors like supply voltage, frequency, driven equipment load etc. on performance of motor. Various energy saving options including replacement of existing motor with energy efficient motor are also discussed.

Chapter 3: Transformer

Chapter 3 highlights the working principle of transformer, construction of transformer, losses in transformer, loading of transformer, calculation of efficiency at different loading factor as well as power factor, calculation of life cycle cost of transformer. It also deals with various energy saving aspects in the transformer.

Chapter 4: HT and LT Cables

In this chapter, construction details, specifications of cables, losses in cables, economic aspects in selecting the cable are discussed.

Chapter 6: Capacitor Bank

Chapter 6 deals with advantages of provision of capacitor banks in electrical network, which results in saving in electricity bill and improvement in power factor. Calculation of KVAR requirement for improving power factor from existing value to target value is also highlighted.

Chapter 7: Compressed Air System

Various energy saving options in compressed air system are highlighted in this chapter. Methods to find out the capacity utilization of air compressor and quantification of air leakage in the system are also elaborated.

Chapter 8: Pumps

Chapter 8 mainly deals with centrifugal pumps which are used in majority in industry. Pump characteristics, system curves, sizing of pumps and various energy saving options are highlighted.

Chapter 9: Air Conditioning System

Various energy saving options like calculation of specific energy, coefficient of performance are highlighted in this chapter.

CHAPTER

1

General Aspects of Energy Auditing

1.0 Introduction

Most of us might have come across a letter either from a state government or a central government agency with a direction to get the energy audit done through accredited energy auditors. Now, the question arises, what is energy audit and why government agencies are putting so much emphasis on it?

We are all aware that to run a business, whether it is a big or small industrial or commercial establishment, energy is required. It may be in the form of electricity or in any other form. Most of the energy is obtained from fossil fuel, which is precious and scarce also. Therefore, energy should be used in such a way that wastage is minimized. Energy auditing is required to identify and minimize the wastage of energy so that the use of energy is optimized.

The subject "Energy Auditing" is highly technical and it involves many engineering branches like Electrical, Mechanical, Chemical, etc. For an energy auditor, it is necessary that he should have complete knowledge of the system for which he is going to perform energy auditing. For example, in a process plant energy auditor will come across pumps, motors, compressors, fans and blowers and many other types of equipments. Besides this, a little background of finance as well as project management is also needed. Decision-making skills based on available data is also a requirement for an energy auditor.

1.1 Benefits from Energy Audit

In fact, besides compliance of government agencies requirement, there are tangible and intangible benefits of energy audit. Some of the benefits are listed below:

- Usage of energy is quantified.
- Analysis and identification of losses, in mechanical system, chemical process or in any other system/process.
- Development of energy monitoring system.
- Conservation of energy and reduction in energy cost.

- Optimization of energy cost by appropriate energy mix.
- Control of emission of gases that pollutes atmosphere.
- Use of right equipment for improvement in efficiency.
- Use of right technology.
- Benchmarking values of energy consumption.
- Reusing and recycling of energy.
- Development of preventive maintenance and quality control programmes.

1.2 Requirement to Conduct Energy Audit

A clear cut energy policy and approach of the management towards energy management are two important aspects, which need to be taken care of before proceeding for energy auditing. Management support is very much required and it will be extended only when management has clear cut policies towards energy management.

- The next step is to keep a provision in budget for carrying out energy audit and also for implementation of recommendations of the energy auditor.
- Identification of personnel for coordinating the various activities related with energy auditing.
- Identification of process, plant, equipments for energy audit.
- Deciding methodology for detailed energy audit.
- Prequalification of energy auditors.
- Preparation, issuance of tender documents for hiring of energy auditing firm.
- Hiring of energy auditing firm.

After identification of the plant and the equipments for which energy auditing is to be carried out, methodology for detailed energy audit is finalized.

1.3 Methodology for Energy Audit

A structured methodology is helpful in carrying out the energy audit in a more efficient way. Methodology varies from one organization to another; however, the objectives of energy auditing should be fulfilled whatever methodology may be adopted. It is important that the methodology adopted should be known to the external energy auditor before hand, so that the energy auditor remains aware about the activities, he is supposed to do during the auditing phase. This can be discussed during pre bid discussion. Some of the steps for finalizing the methodology for energy audit are mentioned below:

- Preparation of single line diagram of electrical system. An example of single line diagram is given in Fig. 1.0. Average load on transformers and on various other feeders are required to be mentioned.

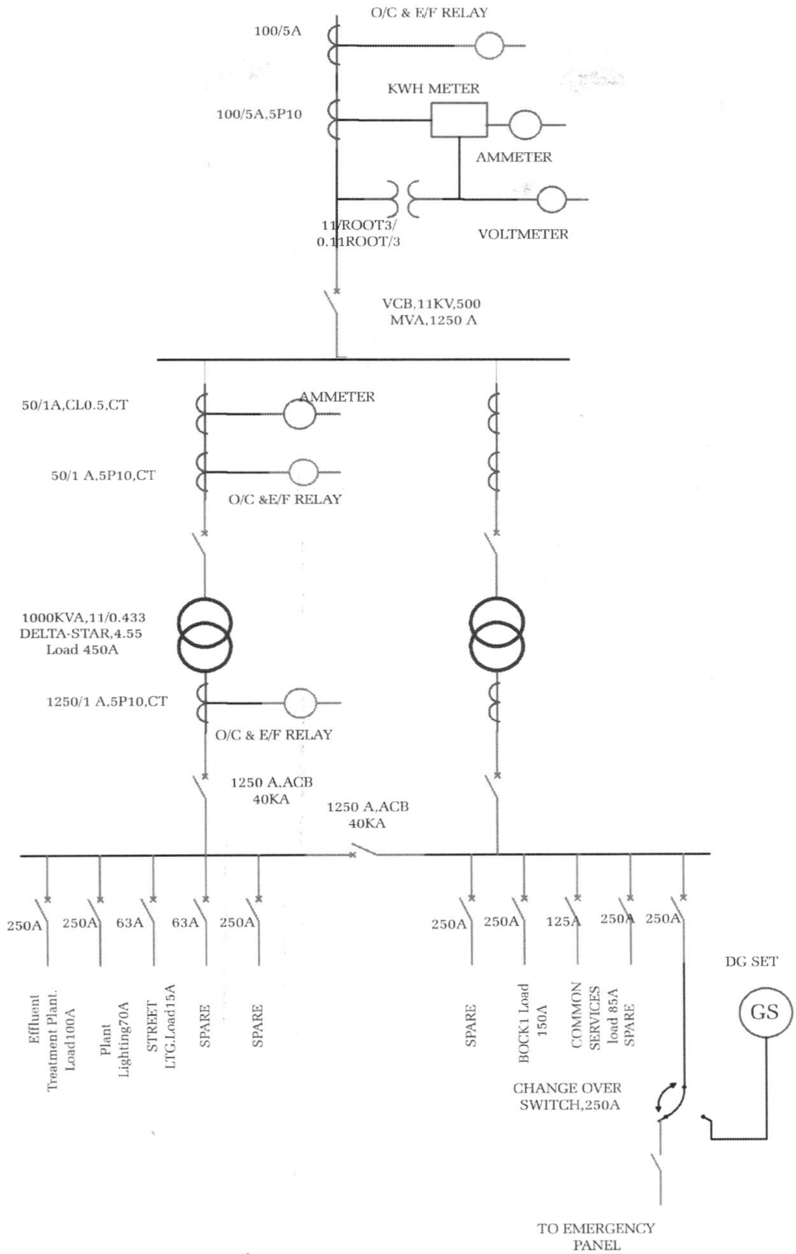

Fig. 1.0 Single line diagram of electrical system

- Collection of baseline data related with energy consumption.
- Study of annual energy bill and energy consumption pattern. Preparation of pie chart, graph etc. of the energy consumption pattern. Identification of energy

consumption per unit of product and also cost of energy per unit of product. Identification of cost of energy in total cost of product. These are all baseline data, which are needed during the detailed analysis of energy consumption.

- A schedule for taking measurements of site is to be prepared in consultation with the engineer-in-charge, so that plant personnel could be spared and deputed along with the auditing team.
- Measurement of various electrical parameters of electrical equipments like motor input power, current, power factor, efficiency at the existing load, load on transformer, existing rating of capacitors banks and existing power factor, loading pattern of existing DG sets, losses in cables and switchgear etc. During this phase, to establish the load variation pattern of the equipments, monitoring of load on continuous basis is required.
- Energy and material balance, energy loss and wastage analysis is to be done.
- Provisions in electricity tariff should be analysed to reduce the power cost like optimization of maximum demand for reduction in maximum demand charges, improvement in power factor for availing rebate in the bill for improved power factor.
- Online recording of load pattern and measurement of parameters of the equipments to evaluate their performance with respect to energy consumption by comparing these parameters with the benchmarked values. If benchmarked values are not established, then these are to be set as per the best practices in the industry and results are to be compared with those benchmarked values. The methods for improving the efficiency are to be suggested. If retrofitting or total replacement of old equipment is recommended, then its cost benefit analysis should also be worked out.
- Presentation of the results obtained by analysis of data, to the top management is an important part of any energy audit. If recommendation for any energy saving scheme is made then the other details like its cost benefit analysis, time required for implementation, technical specification, vendor's address etc. as required should be provided.
- It should also be indicated that the suggested energy saving scheme is new one or retrofitting in the existing system is required. Further, it should be prioritized by short, medium and long-term measure.
- Training programme for general awareness may also be kept in the scope of energy auditor however, this requirement is optional.
- The energy auditor may also study the existing energy monitoring and reporting scheme and if any improvement is required, it may be suggested along with the cost benefit analysis.
- Many times, harmonic study and thermography of the equipments are also kept as part of energy audit programme.

- Sometimes, involvement of energy auditor in implementation of energy saving scheme is also kept as part of energy audit. The decision to involve energy auditor in implementation of energy saving schemes depends on organization to organization. Many organizations have technical people and they are capable to implement the schemes as suggested during energy audit and may not need the services of energy auditor, while some of the organizations need services only on limited basis.

1.4 Preparation of Tender Enquiry Document

After finalization of methodology, the next step is to appoint an energy auditor. Again this procedure varies from organization to organization. For example, in case of private organizations, renowned or accredited energy auditing firms called for discussion and technical and commercial terms are finalized across the table, while in public sector and government organization, an elaborate tendering procedure is adopted. The first step for this is to call for applications for prequalification of the parties. Based on the work experience, availability of resources like tools and tackles, software, technically qualified staffs, financial position etc. parties are prequalified and tender documents are sent to them for competitive bidding. The following are the main clauses, which should be part of enquiry documents.

- Scope of work clause
- Time frame clause
- Secrecy clause
- Methodology clause
- Commercial terms and conditions clause

1.4.1 Scope of Work Clause

The scope of energy auditing should be clearly specified. The information like drawings, documents, test reports and certificates, equipment designed data etc. which will be made available to them during energy auditing phase, should be clearly mentioned. The following information should be a part of enquiry document.

- Specific energy consumption.
- Percentage cost of energy in total cost of product.
- Energy required for per unit production. For example, if electrical energy is the energy source, then units (KWH) of electrical energy required for producing a single unit of product shall be specified.
- Quantity of fuel like gas, coal, furnace oil, etc. required for per unit of production.
- Sources of energy and the total expenditure for the energy in the current year and the past two years.

- A list of electrical and mechanical equipments, for which energy auditing is to be carried out, should be a part of enquiry document. The list should carry information like supply voltage, rating, load factor, operating time in a year, capacity, process detail, electrical network diagram. These details are required so that the prospective energy auditor can have an idea of the quantum of work.
- Enquiry document should clearly specify the list of the documents to be provided to the auditor during the auditing phase and also the documents and drawings which are to be generated by the auditor. For example, a particular motor is working for the last ten years and during this period re-winding of motor has also been carried out due to failure of insulation of motor, then auditor is supposed to carry out the efficiency test on the motor for that particular load at which motor is running and compare the result with the original test certificates. In this way, a new test report is required to be generated by the auditor.

1.4.2 Time Frame Clause

Time of commencement and completion of energy auditing work, submission of draft report, discussion on draft report, and submission of final report is to be specified. Further, time schedule of implementation of scheme as suggested by energy auditor should also be specified. However, this depends on organization to organization, whether they want to fully or partially involve energy auditor for implementation of energy saving scheme.

1.4.3 Secrecy Clause

Energy auditor is not supposed to divulge the data of one organization to another organization , therefore, he is bound by putting a secrecy clause in the enquiry document, that the data provided to energy auditor during the time of energy auditing, shall be kept secret and it should be returned back after completion of the audit.

1.4.4 Methodology Clause

This is an important aspect and already explained under sub-heading "1.3 Methodology for Energy Audit".

1.4.5 Commercial Terms and Conditions Clause

Under this clause, mainly the commercial clauses like schedule of release of payment on the basis of percentage of work done, taxes and duties etc. are specified. For example, it can be mentioned that 70% of payment shall be released on submission of draft report, 10% on submission of final report and the balance payment shall be made on acceptance of report. Sometimes, the energy auditing firms insist on 10%

advance payment. Again, payment terms vary from organization to organization and it should be as per mutually agreed terms and conditions.

1.5 Energy Audit Report

Energy audit report is a very important document and it should be preserved for future reference. The report should be well-structured and should cover all the aspects of energy auditing. In general, the energy audit report should cover the following points:

- Executive summary
- General aspects of energy auditing
- Energy tracking system
- Utilities infrastructure
- Assessments and recommendations
- Operation and maintenance practices

1.5.1 Executive Summary

The first part of the report should be executive summary. In executive summary the following points should be covered.

- Findings regarding the existing energy consumption and the existing energy cost. Percentage of energy cost in total cost of product.
- Variation in the energy consumption from the baseline or benchmarked values.
- Recommendations of schemes and suggestions for saving in energy cost by enhancement in performance of equipments as well as by prevention of wastage of energy, implementation cost of schemes and the payback period. The recommendations may be divided into short-term measure, medium-term measure and long-term measure.
- Benchmarking of energy consumption, equipment-wise, and plant-wise, if benchmarking is not done earlier.
- Improvement in operation and maintenance practices which will directly or indirectly affect the energy consumption pattern.

As far as possible, the above should be depicted in graphical or tabular form.

1.5.2 General Aspects of Energy Auditing

The following points are covered under this heading:

- General description of the organization where audit and its activities has been conducted.

- Audit scope, schedule and methodology.
- A brief description of facilities like electrical system, and other utilities along with system diagram.
- Data collection and investigation.

1.5.3 Energy Tracking System

Data related with utilization of energy with respect to certain time period is collected. Few salient points of energy tracking system are given below:

- Measurement, collection of data with data logging instrument to track the energy consumption pattern of the equipments.
- Establishment of baseline energy usage for all process in KWHR per unit output and any other parameters based on various specified fuel type and feed stock.
- Comparing the present energy usage with benchmarked data.

1.5.4 Utilities Infrastructure

Information related with existing electrical, water, steam, compressed air system are collected. Type, capacity and physical description of equipments, age and condition of the system, and maintenance standard and problem associated with the system are also taken into account.

Further, past data related with studies performed on the system and problems and opportunities identified, should also be collected.

1.5.5 Assessment and Recommendation

In the final and concluding part of the energy audit report, the following points should be highlighted:

- Performance evaluation of electrical system and equipments.
- List of options in term of no cost, medium and high investment cost including payback period.
- Implementation plan for energy saving scheme, time schedule for implementation of schemes, availability of equipments and vendors for the above schemes.

1.6 Conclusion

In this chapter an attempt has been made to have an idea of different phases of energy audit and also about perquisite for the energy audit. We have also dwelt upon the activities to be carried out during the detailed energy audit and finally, structure of energy audit report is also discussed.

1.7 Software

1.7.1 Objective

To prepare a database with details like address, phone no., name of contact person etc. of an organization, industrial establishment where the energy auditing is to be carried out.

1.7.2 Procedure

The information like name of industrial, commercial or any other establishment, product/products manufactured, addresses, name and phone numbers of contact persons etc. is stored under the following head.

- Plant-ID (Plant Identification Number)
- Name
- Product
- Address
- City
- Pin
- Contact person
- Phone
- Fax
- Email

To store the above information, go through the following steps.

- Start the software program, a screen as shown in Fig. 1.1 will appear.
- Click on "Plant Details" of the main menu of the screen, as shown by an arrow in Fig. 1.1.

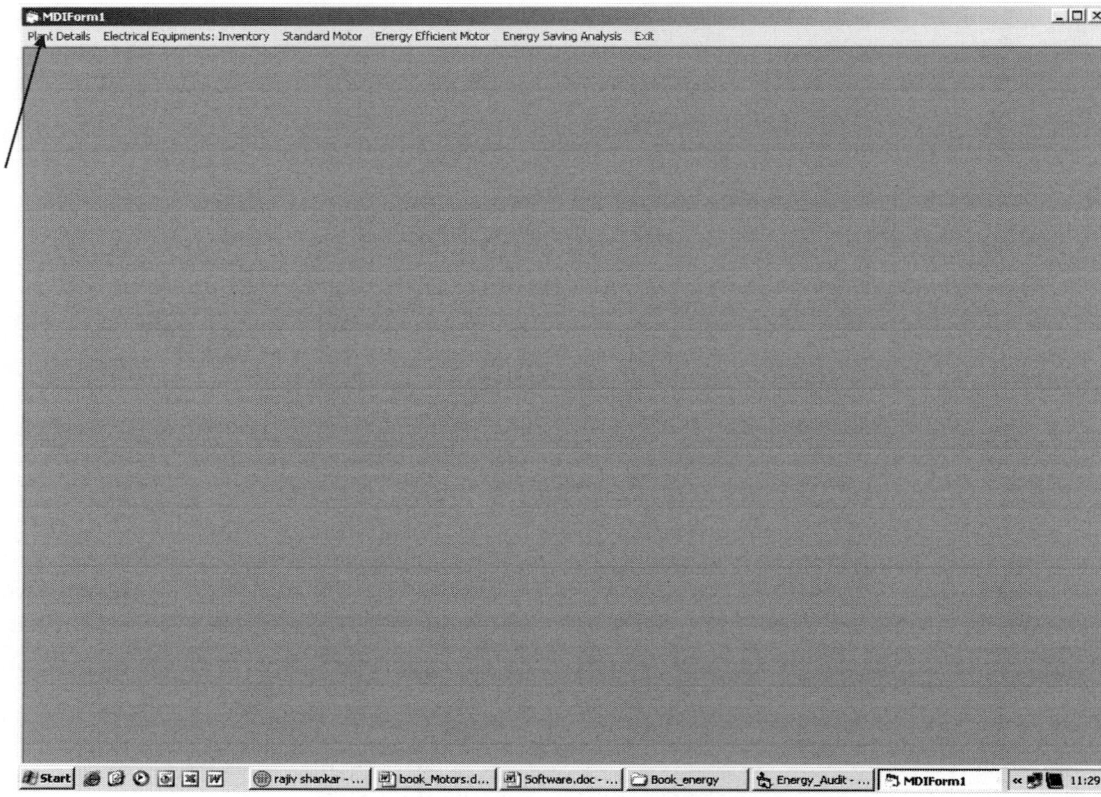

Fig. 1.1 Main screen

- Another screen as shown in Fig. 1.2 opens up.
- For adding new record click on the "New" button and fill the information in the blank boxes. The first box is Plant-ID. This is a plant identification number. It can be any numeric or alpha numeric number. This box should never be left blank and compulsorily be filled for each record.
- After filling up the details in the other boxes, click on "Save" button to save the record.

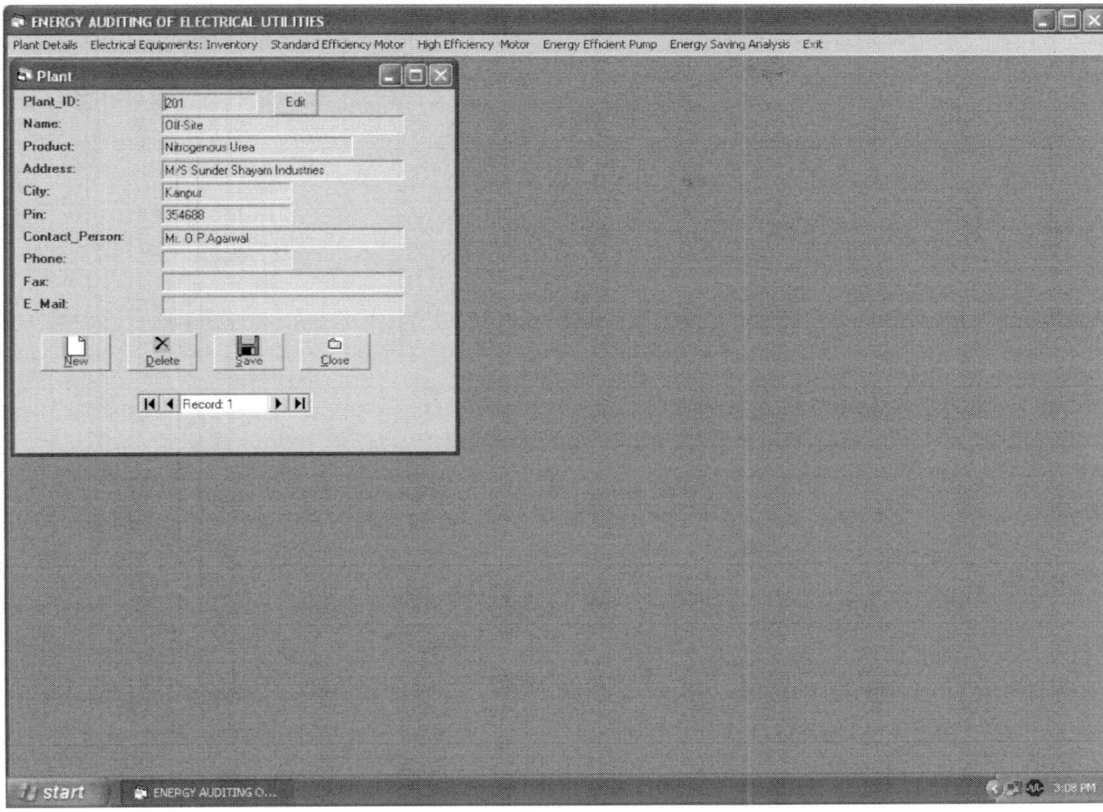

Fig. 1.2 Plant details

- If deletion of a record is required, i.e. deletion of complete information from all the boxes, click on the "Delete" button. Information from all the boxes will be removed and blank boxes will appear.
- The screen can be closed by clicking the "Close" button.
- We can navigate through various records by clicking the navigation bar provided in the bottom of the screen.

When number of records are more and we want to find out a particular record quickly, click on "Find" button, a screen will appear as shown in Fig. 1.3.

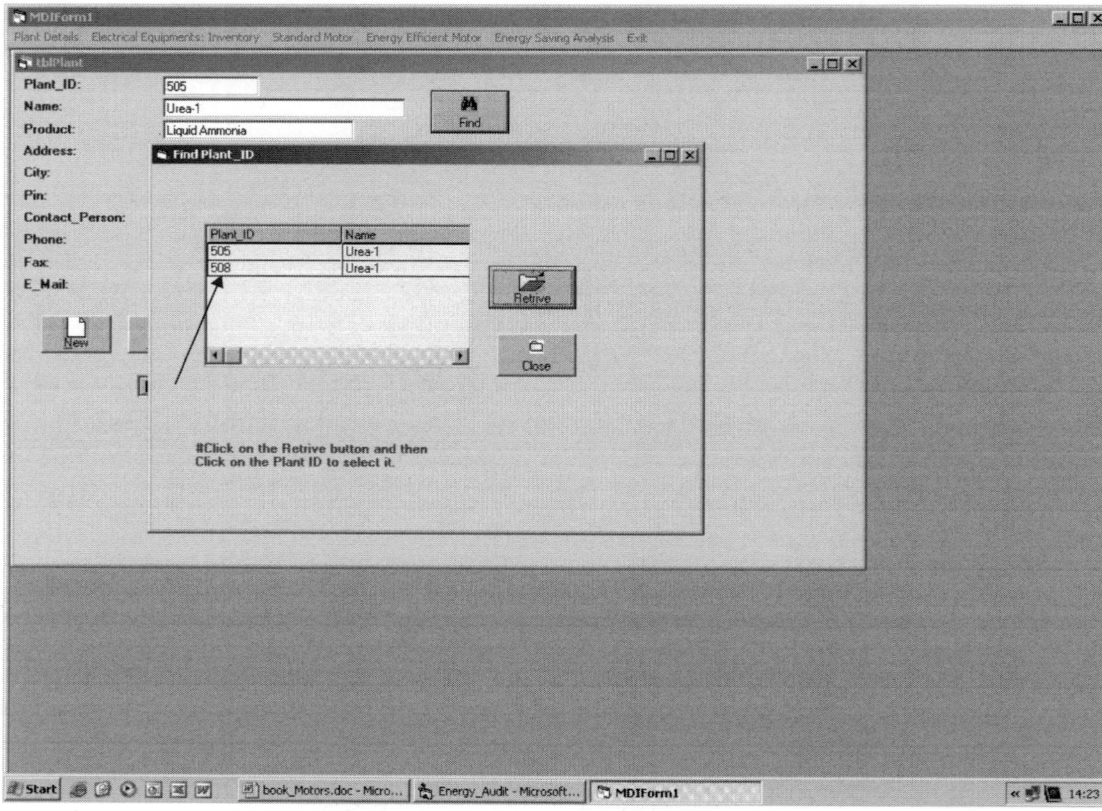

Fig. 1.3 Search screen

This screen consists of two columns namely, Plant-ID (Plant Identification Number) and Plant Name. By clicking on "Retrieve" 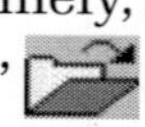 button, the rows and columns will be filled with plant identification number i.e. Plant-ID and name of the plant.

- To edit any record, click on the Plant-ID of that particular plant as shown by an arrow in Fig. 1.3, the search screen will disappear and boxes of main screen will be filled with the records of that particular plant i.e. the clicked record will be the current record and addition or deletion of information can be done. The edited record should be saved by clicking on "Save" button as mentioned earlier.

CHAPTER

2

Squirrel Cage Induction Motor

2.0 Introduction

Squirrel cage induction motors are the driving force of any industry. They cover almost 90% load of an industry. These motors are very rugged, reliable and efficient. The efficiency of any standard squirrel cage induction motor varies from 80% to 90%, while the efficiency of energy efficient motor may be as high as 95%, however, the efficiency depends on KW rating of motor, loading, power quality of the system and various other factors.

Fig. 2.0 Induction motor

Earlier, DC motors and slip ring induction motors were used for speed control and high starting torque applications but with the advancement in technology of variable speed drive, DC and slip ring induction motors are replaced by squirrel cage induction motors and it takes care of all the types of load encountered in an industry.

As compared to DC motor and slip ring induction motor, less maintenance is required in a squirrel cage induction motor. Due to the above mentioned reasons, squirrel cage induction motors are most commonly used in the industry. Therefore, the discussion in this chapter is limited to squirrel cage induction motor.

2.1 Construction of Squirrel Cage Induction Motor

The construction of squirrel cage induction motor is simple. It can be divided into two parts.

- Stator (Stationary Part)
- Rotor (Rotating Part)

The stator consists of cast iron body which secures stampings and it is called Core of Motor. As shown in Fig. 2.1, the core is provided at the inner side of the motor. The core consists of the copper coil of a fixed number of turns in slots. The copper wire is insulated with a film of polyester enamel and impregnated with class F (Thermal class up to 155°C) varnish. These coils are interconnected with each other and finally to the terminal box. The three phase power supply to the motor is fed through the terminal box.

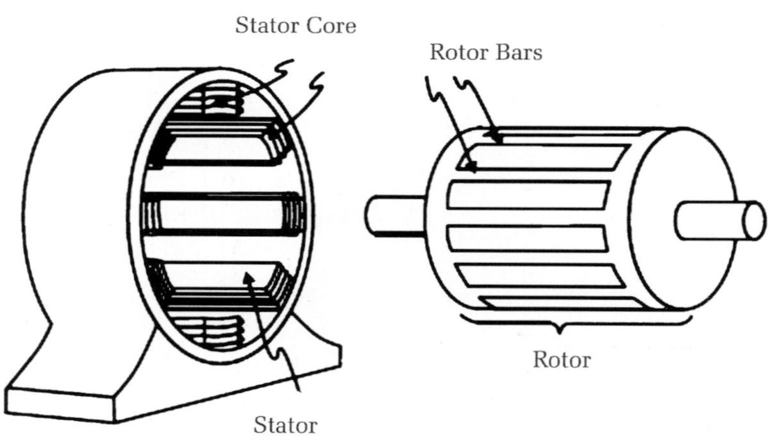

Fig. 2.1 Stator and Rotor of induction motor

The rotor consists of stampings of cold rolled steel secured on the shaft. These stampings form a cage and this cage consist of slots. Copper bars are inserted in the slots and these bars are shorted by providing end rings at both ends of the cage. Refer Fig. 2.1 and Fig. 2.2.

Fig. 2.2 Rotor of induction motor

2.2 Special Design Feature for High Efficiency Motor

Certain changes are incorporated in the design of high efficiency motor. Some of these changes are listed below:

- Copper of higher cross section area is used. The higher cross section area results in lower copper losses due to low resistance.
- The core length is longer in comparison to the standard efficiency motor for reduction in core losses.
- Stampings of low loss (hystersis loss) magnetic material is used.
- The design of stator and rotor is special to reduce to eddy current loss.
- Special fan and fan cover design to reduce the windage loss.

2.3 Operation of Induction Motor

When AC power supply is fed to the squirrel cage induction motor, a rotating magnetic field is produced in the air gap between stator and rotor. The speed of rotation of this field is called synchronous speed and depends on frequency of power supply and number of poles in the winding. The synchronous speed can be calculated by the following formula:

Synchronous speed = 120 × F/P
F = Power supply frequency in Hz
P = Number of poles

The basic principle of working of an induction motor is law of induction and law of interaction. As per law of induction, when we put a coil of N number of turns in an alternating magnetic field, a relative motion between the coil and magnet field takes place. Due to this, the flux linkage with the coil changes and an emf (electro motive force) is induced in the coil. The magnitude of this emf depends on the number of turns in the coil, rate of change of flux linkage and the flux density. If the coil is closed, a current will be induced in the coil and due to interaction of this current and magnetic field in which the coil is lying, a mechanical force perpendicular to both i.e. direction of current and magnetic field is produced. The magnitude of this force depends on density of magnetic field, length of coil and current in the coil. Refer Fig. 2.3.

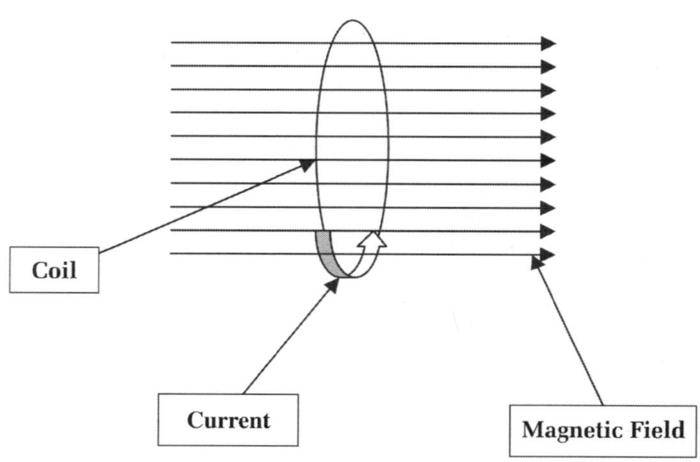

Fig. 2.3 Principle of working of induction motor

If the length of the conductor lying in a magnetic field of B (weber/metre^2) flux density is L (metre) and conductor also carries a current I (ampere) due to induction, then the force on the conductor will be:

F = BIL newton

When a three phase supply is fed to the stator, a magnetic field is produced which rotates between the air gap of stator and rotor at synchronous speed. This rotating magnetic field also moves the rotor and induces an emf in the rotor bars. Since rotor bars are short circuited, it produces a current in the rotor bars. Interaction of rotor current and the rotating magnetic field produces a torque, which rotates the rotor in the same direction in which the magnetic field rotates. The rotor tries to catch the speed of the magnetic field but it never does so. There will always be a difference between the speed of rotor and the magnetic field (which is rotating at synchronous speed). The ratio of this difference in speed and rotor speed is called Slip. The slip is represented by the following formula:

$$S = (Ns\text{-}N)/Ns$$

Where, Ns = Synchronous speed

N = Rotor speed

S = Slip

The rotor speed will be very close to the synchronous speed when the shaft of the motor is not coupled with the driven equipment. This type of condition is called No Load Condition.

When the motor is stationary, the slip will be one and the rotor current frequency will be equal to the mains supply frequency. Since, during stationary condition, the relative movement between the rotating field and the rotor conductor is more, the rotor current will be more and more torque will be produced. As soon as, the rotor gains speed, the relative movement between the rotating field and the rotor conductor reduces and slip also reduces. The rate at which the rotating field cut the rotor conductor also reduces. This results in induced emf in the rotor of less magnitude and therefore, rotor current and torque get reduced.

Now consider a mechanical load is applied on the motor shaft. The rotor will slow down and slip will increase. Emf induced in the rotor will also increase due to increase in relative movement of rotating magnetic field and rotor conductor. The increased emf will produce more rotor current and increase in rotor current will produce more torque. A condition of equilibrium will reach with development of sufficient torque to meet the load requirement. This condition is called full load condition of motor. A slip of 4% or 0.04 in a medium size motor will be sufficient for development of full load torque. Thus, there will not be appreciable change in the speed of the motor to meet the full load condition and the speed of induction motors almost remains constant.

The value of slip varies whenever the rotor comes from stationary state to rotating state. For stationary rotor the slip is 1 and for a rotor rotating at a constant speed the value of slip is less than 1.

The above will be clear from the torque-speed characteristic of the motor.

2.4 Torque-Speed Characteristics

This is one of the most important characteristics which should be taken into account whenever a motor is selected for a particular load. The motor should be capable to produce sufficient torque during starting period as well as during running. Fig. 2.4 shows a typical torque-speed characteristic of an induction motor. The motor has a moderate starting torque and the pull up torque exceeds the full load torque.

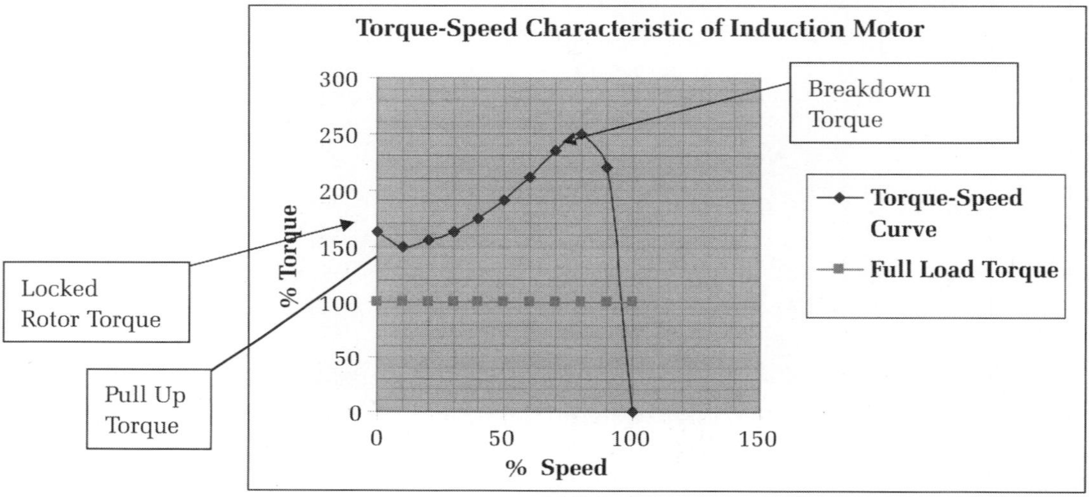

Fig. 2.4 Torque-speed characteristic of motor

The torque-speed characteristic of the motor and driven equipment should be matched. Sometimes, it happens that too much safety margin is kept, due to which oversized motor is selected. This results in inefficient operation of motor and wastage of energy. To improve the efficiency of oversized motor, one of the options is to reduce the supply voltage. Reduction in supply voltage will result in reduction in torque also, but it will not do any harm since the motor is already oversized and that much torque is not required. Reducing the supply voltage will result in reduction in core losses and improvement in efficiency and power factor of the motor.

During starting of motor, the rotor resistance plays an important part. If the resistance is more, stating torque will be more.

Torque is proportional to square of the supply voltage and if the supply voltage is reduced to 80%, torque will be reduced to 64%. Many times, the technical specifications of the motor states that motor should be able to start and accelerate to its rated speed at 80% of supply voltage. This means that motor should develop sufficient torque to cross the pull out torque and to reach the rated speed even at 80% of its supply voltage. If the voltage is further reduced than a condition will come when motor will not be able to overcome the pull out torque and it will not reach to its rated speed.

Under such condition, it will continuously rotate at a speed which is much lower than the rated speed. It will draw a higher current (approximately two times the rated current) and finally, the protective device will operate to stop the motor.

The value of slip at full load ranges 0.02 to 0.05. Torque-slip curve has a steep straight line near the zero slip and it is evident from the curve that a small change in the slip causes a large change in the torque.

2.4.1 Types of Load

The requirement of power from the motor depends on the load-torque characteristic of the driven equipment. For example, the power required is proportionate to the cube of speed for loads like fans, centrifugal pumps etc. Therefore, whenever speed increases more power is required from the motor. The classification of different types of load is given below.

Two types of loads are generally encountered in an industry.

- Constant load torque
- Variable load torque

Lift, crane and reciprocating pumps fall in constant load torque category. High torque is required due to high inertia and it remains constant throughout the speed range. Motor should be capable to provide high torque during starting.

The other types of loads are centrifugal pumps, fans, compressors, mixers etc. where the torque is proportional to square of the speed. With the increase in speed, torque and power requirement increases. This is the reason that a centrifugal pump is started with the suction value fully open and the discharge valve partially close, so that the motor could accelerate to its rated speed with less torque and power. When motor reaches to its rated speed, the discharge valve is opened which causes additional load on the motor. As mentioned earlier, the increase in load causes a little change in slip and since the torque-speed characteristic is a straight line, little change in speed results in development of large amount of torque. Similarly, centrifugal compressor is loaded when motor reaches to its rated speed.

2.5 Operating Parameters of Motor

2.5.1 Motor Power Factor

The induction motor load is a reactive load and the power factor is lagging. The lightly loaded motor has more lagging power factor than a loaded motor. Both active and reactive power is required for operation of a motor. Magnetic field is produced by the reactive power while active power produces the torque for real work. A lightly loaded motor has more reactive power component and therefore, it has poor power factor. If the motor is directly connected with the grid, this reactive power will be supplied by the grid resulting in loading of the grid. Therefore, utility companies insist upon to provide capacitor banks for power factor compensation. If a capacitor bank is provided locally, the reactive power requirement will be fulfilled by the capacitor bank and the installation of utility company for transmitting power to the motor will be relieved.

Therefore, it is always recommended to provide capacitor banks for motors. The matter related with power factor improvement is discussed in more detail in Chapter 6.

2.5.2 Motor Efficiency

Efficiency is another important factor which should be taken into account while purchasing a motor. The efficiency of any motor can be defined as the ratio of mechanical power output to electrical power input. All motors available in the market are not of same efficiency. The name plate of an energy efficiency motor, indicates its efficiency at rated voltage, speed and load. Efficiency varies with loading of the motor. Efficiency of a lightly loaded motor will be less than a fully loaded motor. It also depends on other parameters like, operating voltage, frequency etc.

IS:12615:2004 specifies the efficiency of motors for two efficiency classes of motors. One is improved efficiency (eff2) and another is high efficiency (eff1) class. Refer Table 2.1 to Table 2.5, where efficiency for different rating of motors are shown.

Table 2.1 Two Pole Energy Efficient Motor

KW	*Improved Efficiency (eff2)*	*High Efficiency (eff1)*
0.37	66.0	70.2
0.55	70.0	74.0
0.75	73.0	77.0
1.1	76.2	82.8
1.5	78.5	84.1
2.2	81.0	85.6
3.7	84.0	87.5
5.5	85.7	88.6
7.5	87.0	89.5
9.3	87.7	90.0
11.0	88.4	90.5
15.0	89.4	91.3
18.5	90.0	91.8
22.0	90.5	92.2

30.0	91.4	92.9
37.0	92.0	93.3
45.0	92.5	93.7
55.0	93.0	94.0
75.0	93.6	94.6
90.0	93.9	95.0
110.0	94.0	95.0
125.0	94.5	95.3
132.0	94.5	95.3
160.0	94.8	95.5

Table 2.2 Four Pole Energy Efficient Motor

KW	*Improved Efficiency (eff2)*	*High Efficiency (eff1)*
0.37	66.0	73.0
0.55	70.0	78.0
0.75	73.0	82.5
1.1	76.2	83.8
1.5	78.5	85.0
2.2	81.0	86.4
3.7	84.0	88.3
5.5	85.7	89.2
7.5	87.0	90.1
9.3	87.7	90.5
11.0	88.4	91.0
15.0	89.4	91.8
18.5	90.0	92.2
22.0	90.5	92.6
30.0	91.4	93.2
37.0	92.0	93.6
45.0	92.5	93.9
55.0	93.0	94.2
75.0	93.6	94.7
90.0	93.9	95.0
110.0	94.4	95.2
125.0	94.7	95.5

132.0	94.7	95.5
160.0	95.0	95.8

Table 2.3 Six Pole Energy Efficient Motor

KW	*Improved Efficiency (eff2)*	*High Efficiency (eff1)*
0.37	65.0	69.4
0.55	68.0	72.0
0.75	71.0	74.6
1.1	74.0	77.3
1.5	76.0	79.6
2.2	79.0	82.2
3.7	82.5	85.1
5.5	84.5	86.8
7.5	86.0	88.1
9.3	87.0	89.3
11.0	87.5	89.7
15.0	88.5	90.5
18.5	89.5	91.3
22.0	90.0	91.8
30.0	91.0	92.6
37.0	91.5	93.0
45.0	92.0	93.4
55.0	92.5	93.8
75.0	93.0	94.2
90.0	93.3	94.5
110.0	93.5	94.6
132.0	93.8	94.9

Table 2.4 Eight Pole Energy Efficient Motor

KW	*Improved Efficiency (eff2)*	*High Efficiency (eff1)*
0.37	62.0	66.8
0.55	67.0	71.1
0.75	70.0	73.8
1.1	72.0	76.2
1.5	74.0	77.9

2.2	77.0	80.5
3.7	80.0	83.0
5.5	82.5	85.1
7.5	84.0	86.4
9.3	85.0	87.3
11.0	86.0	88.1
15.0	87.0	89.0
18.5	88.0	89.8
22.0	88.5	90.2
30.0	90.0	91.5
37.0	90.5	91.9
45.0	91.0	92.4
55.0	91.5	92.8
75.0	92.3	93.5
90.0	92.8	93.9
110.0	93.3	94.3

Table 2.5 Four Pole Standard Motor, Degree of Protection IP44 or Superior. IS:8789

KW	*Efficiency*
0.37	64.0
0.55	69.0
0.75	70.0
1.1	73.0
1.5	76.0
2.2	79.0
3.7	83.0
5.5	84.0
7.5	85.0
9.3	85.5
11.0	85.5
15.0	86.0
18.5	87.0
22	87.5
30	88.5
37	88.5

The price of an energy efficient motor is 15% to 30% more than the standard efficiency motors. At this point, question may arise that for a particular application whether we should purchase an energy efficient motor or a standard motor whose price is comparatively less than the energy efficient motor. To arrive at certain conclusion, it is necessary to do a little bit analysis to find out the ownership cost of the motor. Ownership cost is that cost which is incurred for operation of the motor. To calculate the ownership cost software attached with the book will be helpful. We will discuss in more details about the ownership cost in later sections. A standard motor of medium range operates with typical efficiencies range of 85% to 92 %, while an energy efficiency motor operates in 92% to 95% efficiency range.

The standard formula for calculating efficiency is as given below:

$$\text{Efficiency} = \frac{\text{Output power}}{(\text{Output power} + \text{Motor losses})}$$

As mentioned in the formula, the efficiency of a motor depends on the losses which take place in the motor when motor is in operation.

2.6 Losses

Let's have a look on the type of losses which take place in a motor and also how these losses are reduced after making certain changes in the construction of an energy efficient motor?

The losses can be divided into two parts.

- Constant losses or sometimes these losses are called core losses.
- Variable losses

2.6.1 Core Losses

Core losses consist of hystersis losses and eddy current losses. The flux density in the magnetic structure is a major factor that influences the magnetic core losses. The magnetic core losses can be reduced by increasing the length of the magnetic structure and therefore, the flux density gets reduced. Thus, the core in energy efficiency motor is lengthened. This increases the weight of the motor. However, per unit of weight, the magnetic losses will get reduced.

The magnetic core losses can also be reduced by using steel containing silicon up to 4% in the laminations. The better magnetic property reduces the losses by 10 to 25%. On the other hand, the disadvantage is that the permeability will be low and this will increase the magnetizing current requirement resulting in decrease in power factor.

Eddy current losses can be reduced by using thinner laminations and also by ensuring adequate insulation between the laminations. In this way the path of flow of eddy currents get reduced and flow of current is also reduced.

2.6.2 Windage and Friction Losses

This loss is due to friction in bearing of the motor. The windage loss in the motor takes place due to power consumed by the ventilation fan and other rotating elements of the motor. The friction loss in the bearing depends on the bearing size, type, speed of the motor, load on the motor and the lubrication used. The friction loss is fixed for a particular design of motor and comparatively small in percentage of the total motor losses. The windage loss depends on design of ventilation fan and the amount of ventilation required for removing the heat produced due to various other losses. The reduction in other losses like copper losses, core losses and improved design of fan reduces the windage losses.

2.6.3 Copper Losses in Stator and Rotor

Copper losses constitute almost 50% of total losses in a motor. Stator copper losses make up to 66% of total copper losses in the motor. The reduction in copper losses results in reduction in overall losses and increase in efficiency. The quantity of copper and cross section area of copper wire is more in an energy efficient motor so that the resistance of stator winding gets reduced. The stator slots in an energy efficient motor can accommodate approximately 20% more copper in comparison of a standard motor.

Rotor current depends on the slip and sometimes rotor losses are also called slip losses. If slip is reduced, rotor losses will also get reduced. This is done by increasing the mass of the rotor conductor i.e. rotor bars and end rings.

Copper of higher conductivity is used to reduce the losses in the rotor. Some of the motors also have die cast aluminium bars resulting in reduction in slip, but aluminium has less conductivity than copper. This characteristic of energy efficient motor i.e. low slip and high speed should be taken into account while considering a replacement of standard efficiency motor with an energy efficient motor. In some applications like centrifugal pumps, compressors and fans, power is proportional to cube of the speed. It means when speed increases, power requirement also increases and instead of saving power, power consumption increases. A proper analysis should be done before making a decision for replacement of a standard motor by an energy efficient motor.

2.6.4 Stray Load Losses

Stray load losses are the losses due to leakage fluxes induced by load currents and increase with load. These losses are difficult to determine by direct measurement or calculation. Some of the factors which influence the stray load losses are:

- Stator winding design
- Ratio of air gap length to rotor slot opening

- Ratio of number of rotor slots to stator slots
- Air gap flux density

Stray load losses can be reduced by careful design of the motor.
A typical distribution of losses is given in Table 2.6.

Table 2.6 Typical Distribution of Losses

Stator copper loss	38%
Rotor copper loss	19%
Core loss	18%
Friction and windage loss	12%
Stray load loss	13%

2.7 Measurement of Efficiency

Two types of terminology are used for efficiency of motor.

- Nominal Efficiency:
  This efficiency is the efficiency obtained by testing a sample motor from a batch of motors of same specification.
- Guaranteed Efficiency:
  Guaranteed efficiency is that efficiency for which manufacturer gives a commitment that the efficiency of the motor supplied by them will be equal to or more than the guaranteed by them.

2.7.1 Standards

Following is the list of some of the international standards for testing of efficiency.

- IEEE-112 (Institution of Electronic and Electrical Engineers)
- IEC-34-2 (International Electro-technical Commission)
- JEC-37 (Japanese Electro-technical Committee)
- BS-269 (British Standards)
- CSA-390 (Canadian Standards Association)
- IS:4889 Methods of determination of efficiency of rotating machine (Bureau of Indian Standard)

The efficiency when tested according to different standard will vary to several percentages, since the methodology of measuring efficiency and the assumptions differ in different standards. For example, according to IEC standard assumption is made that the stray load losses are fixed at 0.5 percentage of input power, while the JEC standard assumes that there are no stray load losses.

IS:8789 specifies the values of nominal efficiency, full load speed, full load current, break away torque in percentage of full load torque of a standard motor (up to 37KW).

Two types of efficiency classes are mentioned in IS:12615. Refer Table 2.1 to 2.4

- Eff2: Improved efficiency
- Eff1: Higher efficiency

The Indian Standards specify determination of efficiency by summation of losses method, which is described as below.

2.7.2 Measurement of Resistance

The stator winding resistance should be measured by voltage drop of ammeter-voltmeter method or by bridge method, where a comparison is made between the known resistance and the resistance which is under measurement. Since, the value of resistance changes with change in temperature, the temperature of the winding shall also be measured. If the motor is at standstill for a longer time, then the surrounding temperature may be considered as winding temperature. Correction for temperature in winding resistance may be done by using the following formula:

$R_2 = (235 + t_2) \times R_1/(235 + t_1)$

Where R_2 is unknown resistance at temperature t_2°C and R_1 is measured resistance at temperature t_1°C.

The above formula is valid if the winding is of copper, if winding is of aluminium, 235 should be replaced with 225.

The value of resistance should be corrected for a value of operating temperature at full load of the motor. The typical value is 75°C.

Table: 2.7 Insulation Class and the Reference Temperature

Class of Insulation	*Reference Temperature °C*
A, E	75
B	95
F	115
H	130

2.7.3 Slip Measurement

Slip is most important parameter in determination of efficiency. Measurement of full load speed by tachometer and finding out slip by subtracting it from synchronous speed is not recommended by the Indian Standard. IS:4889 recommends for direct measurement of slip by one of the following methods.

- Stroboscope
- Slip coil
- Magnet method

The most common method used for measurement of slip is by using stroboscope. In this method on one end of the shaft a single black radial line is painted against the white background. The slip is measured by counting the backward motion of the black line for a given period of time.

2.7.4 Loading of Motor

The motor is loaded by dynamometer and output may be calculate by the following formula:

KW = T × RPM/974

Where T is torque in Kgm

RPM is the speed in rotation per minute

If the dynamometer is not available then a calibrated generator can be used for loading of motor.

2.7.5 No Load Test

This test is conducted to find out the following points:

- No load current
- Core losses
- Friction and windage losses

The motor is run on no load at rated voltage, frequency. Input power, voltage and current are noted. The input power is sum of core losses, copper (i^2R) loss in stator and friction and windage losses. The copper losses (i^2R) are subtracted from the input power and the remaining losses will be sum of friction and windage losses and core losses.

If it is desired to separate the friction and windage losses from the core losses, current and input power are measured at different values of voltage and up to 125% of voltage. Then voltage is reduced up to a point where further reduction in voltage increases the current. The copper (i^2R) losses of stator are deducted from the input power reading and a curve is drawn between the watts and the voltage. The curve line is extended backwards so that it cuts the y-axis at a certain point. This point is the friction and windage losses. If we plot a curve using the value of square of the voltage then a straight line will be obtained and friction and windage losses can be obtained by extending this line backwards.

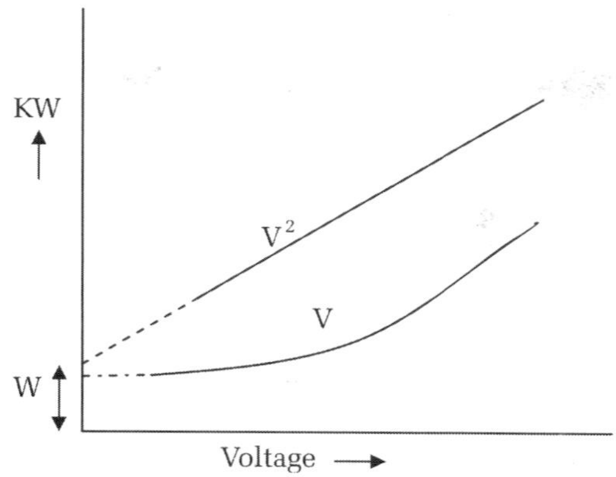

Fig. 2.5 Graph for segregation of friction and windage losses and iron losses

2.8 Determination of Energy Saving

The energy saving depends on the following factors:

- Load at which motor is operating
- Running hours
- Electricity tariff
- Motor size
- Type of motor i.e. standard efficiency motor or energy efficient motor

The following formula is used for calculating the energy saving.

$KW(100/E_1 - 100/E_2)R$

KW is the output at which motor is operating; E_1 is the efficiency of standard motor at the operating load; E_2 is the efficiency of energy efficient motor at the same load; R is the running hours in a year.

If we multiply per unit cost of electrical energy with the saved energy then we will get the total amount of saved energy. It is important to determine the load factor of the motor. Since the efficiency varies with the load, therefore a comparison between efficiency of a standard motor and an energy efficiency motor is made for same load.

2.9 Determination of Load

The load factor can be calculated by the following formula:

$P_1 = KW_1/E_1$

KW_1 = Motor name plate rating

E_1 = Efficiency as per name plate

$P_2 = KW_2/E_2$

KW_2 = Measured output at which motor is working
E_2 = Measured efficiency at which motor is working
Load factor = P_2/P_1

Sometimes, loading of motor is determined just by taking the ratio of line current which is not correct method and likely to give erroneous result.

2.10 Assessment of Economic Feasibility

The cost of energy efficient motor is 15% to 20% more than the cost of standard efficiency motor because of better design and use of higher quality material. Since, the losses will be less; the extra cost can be recovered in a short duration of time. Following methods are used to assess the economic feasibility of investment.

- Simple payback period
- Net present value
- Life cycle cost methodology

The most commonly used method in industrial plant is the simple payback period and most plant engineers desire that the investment which is made in energy efficient motor should be recovered within a period of one to three years through saving in energy. The simple payback period will be the ratio of the cost of new motor and costs of energy saved plus salvage value of the old motor.

The net present value is defined as the difference between the present worth of the saving and the investment made. The residual life of the equipment shall be considered and the following formula is used to calculate the present net worth of the savings.

$$F = P/(1 + r)^N$$

F = Present worth
P = Net annual saving
r = Discount rate
N = Year of saving

r is the yearly rate for fall of value of money which is to be earned in future. After calculating the present value for the predetermined number of year, it is compared with the investment. If the NPV is more than the investment, only then the investment is assumed to be favourable.

The present worth calculation can be simplified for a period of years as per the following equation:

$$F = P\,((1 + r)^N - 1)\,/r(1 + r)^N$$

- Life cycle cost is the ownership cost of a machinery which includes cost of acquisition, operation and maintenance, decommissioning. In a broader term, it is cost of equipment from procurement to disposal. To find out life cycle cost, the present value of total cost of losses is worked out and added with the procurement cost of the equipment. This method is useful when we compare the cost of two equipments of similar parameters and similar operating conditions. Many times, when comparing the life cycle cost, it has been found that though the capital cost of one of the equipment is less in comparison with the other, but its life cycle cost is more. The reason is the cost of losses over its entire life cycle is more.

2.11 Choice of Energy Efficient Motor

Let's assume that a working motor has failed. The choice before the plant engineer is whether to put back existing motor into operation after getting it re-winded or go for purchase of a new standard efficiency motor or an energy efficient motor.

The cheapest option is to get the burned motor re-winded but the question arises whether motor will have the same efficiency as it was before re-winding? In most of the cases, it has been observed that it is difficult to maintain the same efficiency after re-winding and sometimes, efficiency goes down to a large extent, reason being the core losses increase due to change in property of the core material. This change takes place due to heating of core during removal of the old winding. Many times the numbers of turns and even the wire gauge is also not same as it was for the original motor. Decrease in numbers of turn results in increase in starting current and reduction in torque. In such case, where it is economical to get the motor re-winded, plant engineers should take the guarantee from the vendor for the losses of the re-winded motor.

If the motor is old enough and it has been re-winded for two times or more, then it would be better to replace it by the new energy saving motor.

A survey is required to find out number of motors which are older than 15 years and the efficiency of all those motors should be calculated. If motor is found to be working on less efficiency, then the option of replacement of motor with an energy efficient motor should be evaluated. The larger rating motors should be targeted first, because the chances of energy saving with larger rating motor will be more.

2.12 Energy Saving Options in Oversized Motor

Sometimes, the motors of higher rating are chosen for a particular application. A service factor of 1.10 is common in a process plant. For example, load is taken as 7.00KW, after adding service factor it comes out to be 7.7KW and the nearest size available is 11KW. In this way motor is 57% oversized.

Many times, equipment load is not correctly known to the design personnel and motors of higher rating are chosen. Further, to accommodate the future increase in production capacity motors of higher rating are chosen. In some cases, changes in product manufacturing process takes place or sometimes changes in the driven equipment like impeller of pumps, fan blades etc. takes place, due to these changes the lower rating motors are needed in place of existing higher rating motor. The following are the disadvantages of higher rating motors.

- Lower efficiency
- Lower power factor
- Higher motor cost
- Higher controller cost
- Higher installation cost
- Higher operating cost

If a motor is operating at 70% load, it is operating at or near to its maximum efficiency point and much variance in efficiency is not expected with change in load.

If a motor is operating at less than 50% load, it is an oversized motor and needs corrective action. Performance of such motor can be improved by connecting it in star. By connecting in star, the capacity of the motor reduces to one-third, power factor improves and the efficiency also increases.

The second option is to connect a soft starter. The soft starter reduces the voltage when the motor is in running condition. Reduction in voltage results in reduction in torque, improvement in efficiency since the core losses, which are voltage dependent get reduced. By reduction in voltage improvement in power factor also takes place.

The third option is to replace the motor with the new energy efficient motor of smaller size.

In all above cases, the load torque characteristic of the driven equipment should be taken care of. In short, the motor should be capable to accelerate the load from rest to its rated speed within the specified starting time. As mentioned earlier also that fan, blower, centrifugal pump all have variable load torque and the power requirement is directly proportion to cube of the speed.

The full load slip of energy efficient motor is slightly less than the slip of a standard efficiency motor; therefore, the speed of an energy efficiency motor is slightly more than the speed of standard efficiency motor. In case of equipments of variable load torque, like fan, a slight change in speed results in considerable change in the power requirement since power is proportional to cube of speed, while the quantity of air delivery varies linearly with the speed and the end result will be increased energy consumption. Therefore, energy efficient motor should be carefully selected for a load having variable torque.

If it is not possible to get an energy efficient motor of same speed, other methods of energy saving like trimming of impeller, soft starter, converting delta connected motor into star etc. should be explored.

2.13 Affect of Variation of Voltage on Performance of Motor

If motor does not operate on voltage for which it is designed, then its efficiency decreases. Refer Table 2.8, which indicates the affect on operating parameters of motor for variation in voltage. The normal power supply conditions at which motor is designed is:

Voltage	=	415 ± 10%
Frequency	=	50Hz ± 5%
Combined variation	=	± 10%
Ambient temperature	=	50°C

If the variation in supply voltage/frequency is different from the designed values, then motor is to be de-rated as per the manufacturer recommendation.

2.13.1 Over-voltage

As the supply voltage is increased, the magnetizing current increases. At some point, core will get saturated and over heating of core will take place. At about 10% to 15% over voltage both efficiency and power factor decreases significantly. The starting current, starting torque and breakdown torque will increase with increase in voltage.

2.13.2 Under-voltage

If motor is operated at reduced voltage, it will draw increased current to fulfil the torque requirement imposed by the load. The increase in current will lead to increase in copper (I^2R) losses of both stator and rotor.

Table 2.8 Affect of Variation of Voltage on Motor Performance

	110%	*90%*
Starting and Running Torque	Increase 21%	Decrease 19%
Full Load Speed	Increase 1%	Decrease 1.5%
% Slip	Decrease 17%	Increase 23%
Full Load Efficiency	Decrease 1% to 4%	Increase 0.5% to 1%
¾ Load Efficiency	Decrease 2% to 5%	Increase 1% to 5%
½ Load Efficiency	Decrease 4% to 7%	Increase 2% to 4%
Full Load Power Factor	Decrease 10% to 15%	Increase 8% to 10%
¾ Load Power Factor	Decrease 10% to 15%	Increase 10% to 12%
½ Load Power Factor	Decrease 10% to 15%	Increase 4% to 5%
Starting Current	Increase 10 to 12%	Decrease 10 to 12%
Full Load Current	Increase 2% to 11%	Increase 6% to11%

If we compare the efficiencies of motors having different speed and enclosure, then the efficiency of higher speed motor will be slightly more than the lower speed motor. Similarly, the efficiency of motor with open enclosure will be slightly higher than the totally enclosed motor.

2.14 Effect on Efficiency Due to Variation in Load

The energy efficient motor works well on part load also. The efficiency curve of energy efficient motor is almost flat beyond 60% load. It means that no appreciable change in efficiency of the motor takes place beyond 60% of load. Therefore, energy efficient motor is most suitable where load varies 60% to 100%. Refer Fig. 2.6 where a comparison of efficiency is made between standard and energy efficient motor.

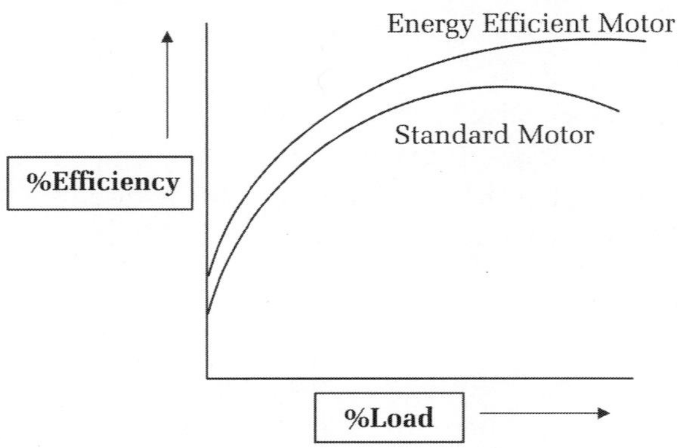

Fig. 2.6 Comparison of efficiency of standard and energy efficient motor of 20KW rating

2.15 Power Factor

Any electrical equipment draws two type of power. One is reactive power, which is required for generating magnetic field and the other is active power which produces the real work. The total power is the vector sum of both active and reactive power. The power factor is the ratio of active to total power. A lightly loaded motor has low power factor and electricity board imposes penalty, if the power factor falls down below a certain limit. This subject has been discussed in more detail in Chapter 6.

2.16 Unbalance Phase Voltage

The unbalance in phase voltages is not a normal condition. If the unbalanced voltage is fed to the motor, it will cause torque pulsation, vibration and increase in stress. The motor losses will increase and heating will take place.

The unbalance in voltage is due to the following reasons:

- Unevenly distributed single phase load on the same power system.
- Unidentified single phase to ground fault.
- Unbalance in supply from the electricity board.
- Faulty operation of automatic power factor correction equipment.

1% of voltage unbalance will increase the motor losses by 5%. In Fig. 2.7 curve showing the effect of voltage variation on voltage losses.

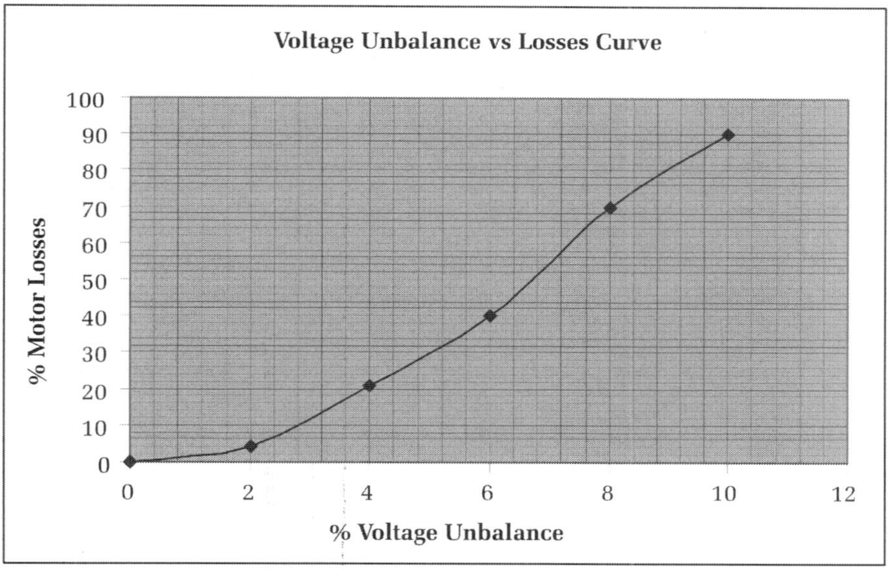

Fig. 2.7 Voltage unbalance vs Motor losses

2.17 Insulation System

The failure of motor takes place due to failure of winding insulation. Most probable reason of failure of insulation is thermal degrading. Generally, low voltage motors fail due to inter-turn short circuiting since no extra insulation is provided between the turns of a coil except varnish coating on copper wire. Inter-turn fault subsequently develops into phase to ground fault or phase to phase fault. Phase to ground fault damages the core also. Motor is protected by the HRC fuses and thermal overload relay. Both of these protections do not give adequate protection for ground fault. The ground fault does not generate sufficient current for blowing of fuses and the thermal relays take time to operate. Therefore, it is beneficial to provide a separate ground fault relay with CBCT (Core Balance Current Transformer) to protect the motor from ground fault.

2.18 Service Factor

Service factor represents the margin in motor capacity. For example, if service factor is 1.15, motor should be capable to operate for 115% increase in load. Sometimes, while developing specifications, it is specified that the insulation class of the motor will be "F" and temperature rise will be limited to insulation class "B". To fulfil this requirement manufacturer supplies motor after de-rating of one size higher motor to keep temperature rise limited to class "B". Such choice leads to operation of motor with poor power factor and poor efficiency.

2.19 Motor Improvement Programme

As mentioned earlier also that induction motors consist of 90% load in an industry, therefore energy consumed by the motors should be carefully monitored. To track the energy consumption, a list of all the motor should be prepared along with the history of motors. History details shall consist of maintenance records of the motors such as re-winding records, replacement of bearing record, date of installation and commissioning of motor, so the age of the motor could be identified. The record like operating load and the efficiency shall also be kept.

In nut shell complete operating and maintenance data is required to make a decision for replacement of existing motor with energy efficient motor. It is recommended to replace the old re-winded and less efficient motors from the batches because manufacturer may pass a good amount in discount if motors sufficient quantities are procured.

2.20 Energy Conservation Building Code (ECBC)

Energy Conservation Building Code (ECBC) has been developed by IIEC, International Institute of Energy Conservation under contract with USAID, United States Agencies for International Development as a part of energy conservation project providing support to Bureau of Energy Conservation action plan. ECBC is mandatory for all commercial buildings or building complex whose connected load is 500KW or more or the contract demand is 600KVA or more. This code is also mandatory for the building complexes having conditioned area 1000 sq.m or more. This code specifies the efficiency requirement for the motors used for various services in the building. Some of the requirements, as specified in the building code are highlighted below:

- All permanently wired poly-phase motors of 0.375KW or more serving the building and expected to operate for more than 1500 hours per year and all permanently wired poly-phase motors of 50KW or more serving the building and expected to operate for 500 hours per year should have a minimum acceptable nominal motor full load efficiency as shown in Table 2.9.

Table 2.9 Minimum Acceptable Motor Efficiency

Motor Size (KW)	*Efficiency% 2Pole*	*Efficiency% 4Pole*
1.1	82.2	83.8
1.5	84.1	85.0
2.2	85.6	86.4
3.0	86.7	87.4
4.0	87.6	88.3
5.5	88.5	89.2
7.5	89.5	90.1
11.0	90.6	91.0
15.0	91.3	91.8
18.5	91.8	92.2
22.0	92.2	92.6
30.0	92.9	93.2
37.0	93.3	93.6
45.0	93.7	93.9
55.0	94.0	94.2
75.0	94.6	94.7

- Motor which is not mentioned in the table shall have efficiency greater than the next KW rating motor in the list.
- Motor rating shall not exceed more than the 200% of the calculated maximum load being served.
- Motor name plate shall indicate the nominal full load efficiency and also full load power factor.
- Motor user should insist on proper re-winding practices for any rewound motor. If proper re-winding practices cannot be assured, the damaged motor should be replaced with the new energy efficient motor rather than suffering from the significant efficiency penalty associated with typical re-winding practices.
- Certificates shall be obtained and kept in record indicating the motor efficiency. Whenever, a motor is rewound, appropriate measures shall be taken so that the core characteristics of the motor is not lost due to thermal and mechanical stress during removal of damaged parts. After re-winding, a new efficiency test shall be performed and a similar record shall be maintained.

2.21 Conclusion

A large chunk of energy is utilized by the induction motors in any industry. Since motor load is a continuous load, it is utmost requirement that the motor should consume electrical energy in a most efficient manner. The oversized, less efficient motor should be identified and necessary action should be taken on priority basis for improvement in energy consumption. Various types of losses encountered in the motor and the techniques for reduction of these losses are explained in this chapter.

2.22 Software

2.22.1 Objective

- To create a data bank of existing motors.
- To create of data bank of new standard efficiency motors and high efficiency motors.
    - The data bank will have facilities like search, addition, deletion, and editing of records.
    - Calculation of efficiency of the existing motors after field measurement of various electrical parameters.
    - Retrieval of data based on query generation from the data bank.
    - Analysis of various options for saving of energy.
    - Calculation of simple payback period.
    - Generation of printout of various analysis.

In general, the software would help us in identifying the inefficient motors, performing cost benefit analysis for taking decision, so that corrective action could be taken to optimize the cost of energy.

2.22.2 Procedure

As mentioned earlier the first step is to develop a data bank for all those motors available in the plant for which energy auditing is to be performed. The procedure is same as mentioned in Chapter 1.

Click on "Electrical Equipments: Inventory" of the main menu, a submenu will appear. Click on "Motors" of submenu as shown by an arrow in Fig. 2.8.

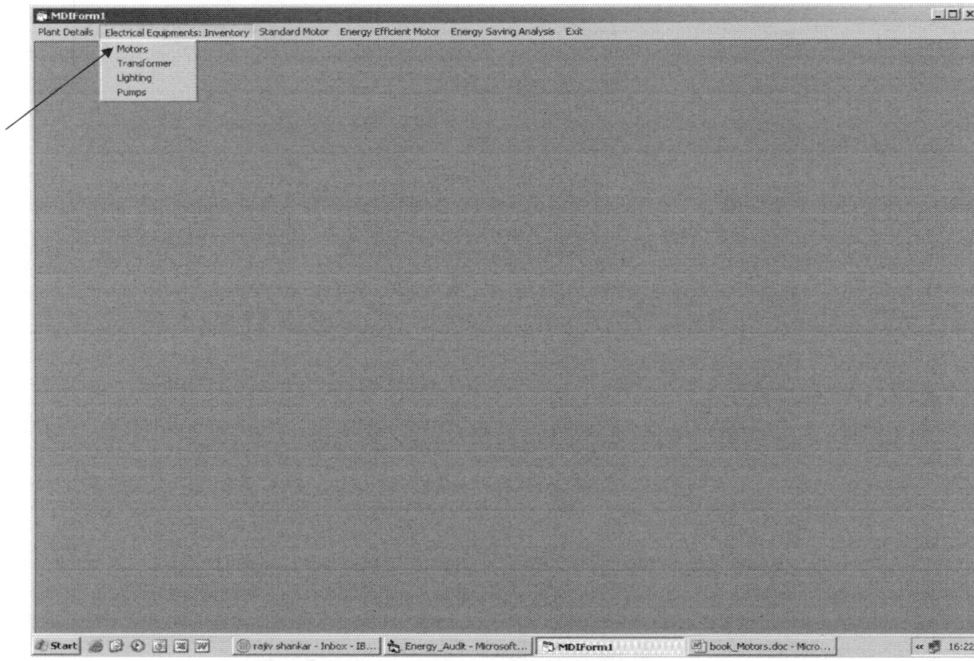

Fig. 2.8 Main screen with submenu

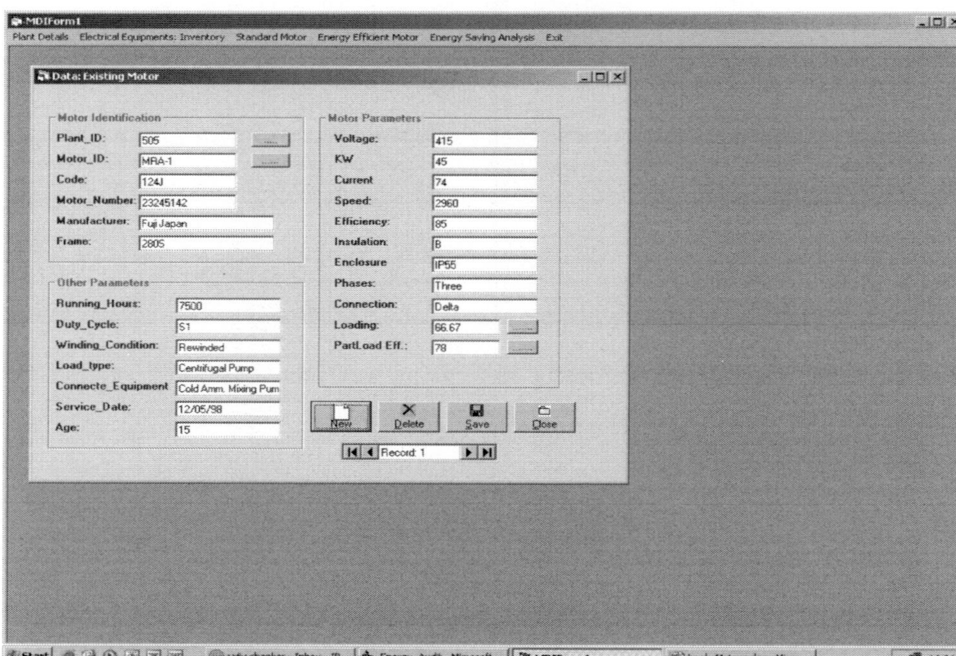

Fig. 2.9 Screen for entry of details of plant motors

- On clicking the "Motors", a screen will appear as shown in Fig. 2.9.
- The first information to be filled in Plant-ID. The Plant-ID should be of that plant to which the motor belongs. If, number of plants are more, it becomes difficult to remember the Plant-ID. Plant-ID can be obtained by clicking on the button provided on the right hand side of Plant-ID box, as shown by an arrow in Fig. 2.10. On clicking the button, a Plant-ID screen will appear, as shown in Fig. 2.11.

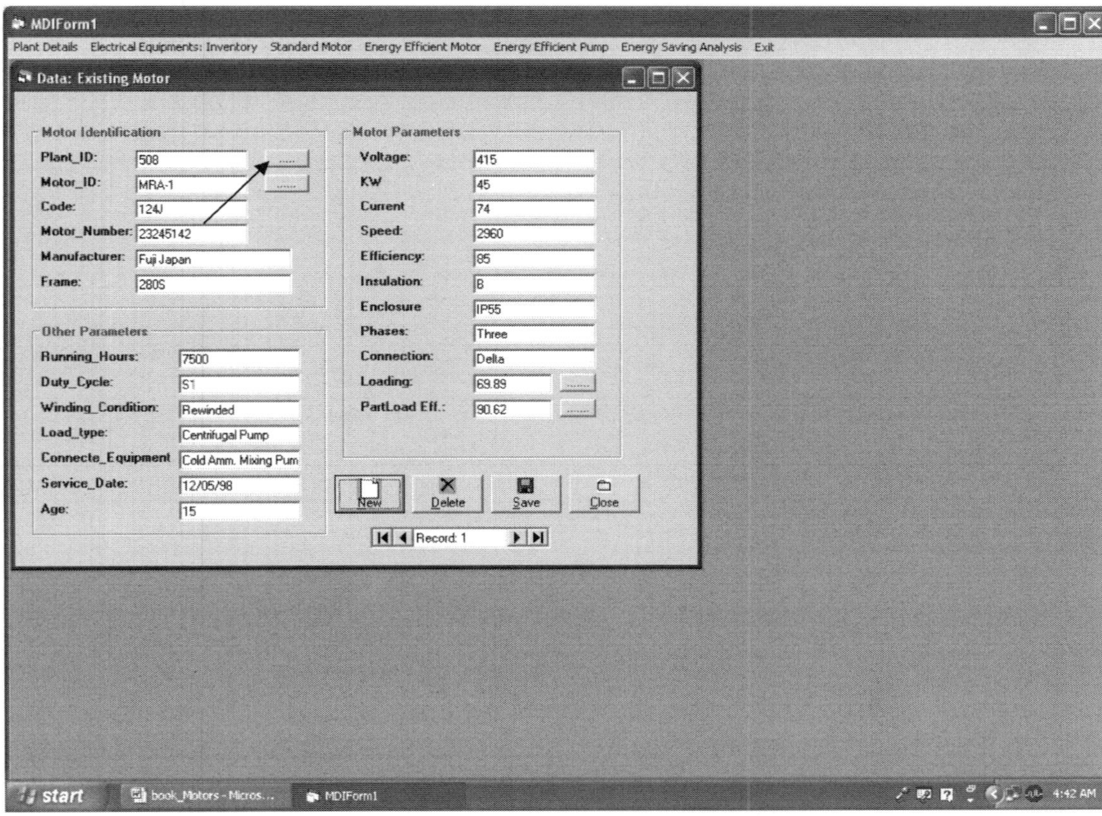

Fig. 2.10 Screen for plant motors

- Click on "Retrieve" button provided in Plant-ID screen, as shown by an arrow in Fig. 2.11, to retrieve the plant name and Plant-ID. As soon as "Retrieve" button is clicked, the blank table is filled up with all the records of Plant-ID and plant name as available in the data bank.
- Copy the Plant-ID in the Plant-ID box of main motor screen. Close the Plant-ID screen by clicking on "Close" button provided in Plant-ID screen.

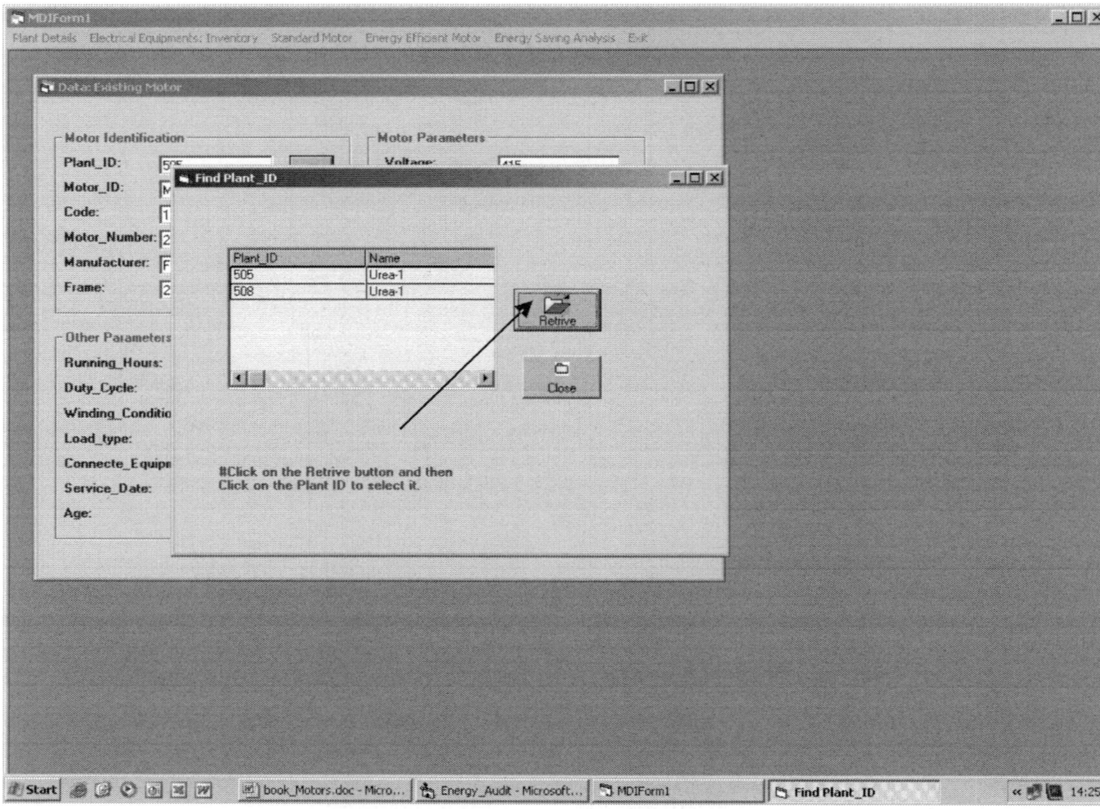

Fig. 2.11 Plant-ID screen

- The second entry is of Motor-ID (Motor Identification Number). Any numerical or alpha numerical number may be entered. However, logic should be applied for generating motor identification number. For example, if motor, for which data is to be entered, belongs to Ammonia Plant-2, Motor-ID (Motor Identification Number) can be MRA-2. The first two letters MR indicate the equipment in motor and A-2 indicates that it belongs to Ammonia Plant-2. An unique number for that particular motor can also be provided like "11". In this way, the Motor-ID will be MRA-211. There may be many other ways and logics for developing the identification number. Care should be taken that two motors should not have same motor identification number. Each motor should have a separate and unique motor identification number.
- One more button has been provided on the right hand side of the Motor-ID box of the main motor screen as shown in Fig. 2.12 by an arrow. This button is for editing of motor records. The procedure is similar as mentioned in Chapter 1 for editing of Plant-ID records. On clicking the "Editing" button, a screen will appear as shown in Fig. 2.13.

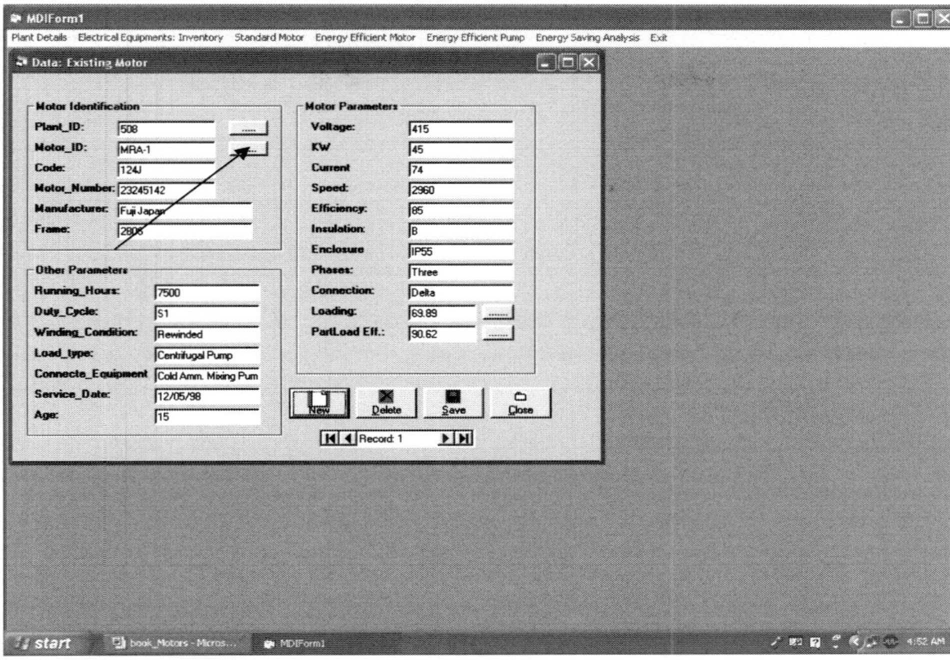

Fig. 2.12 Screen for plant motors

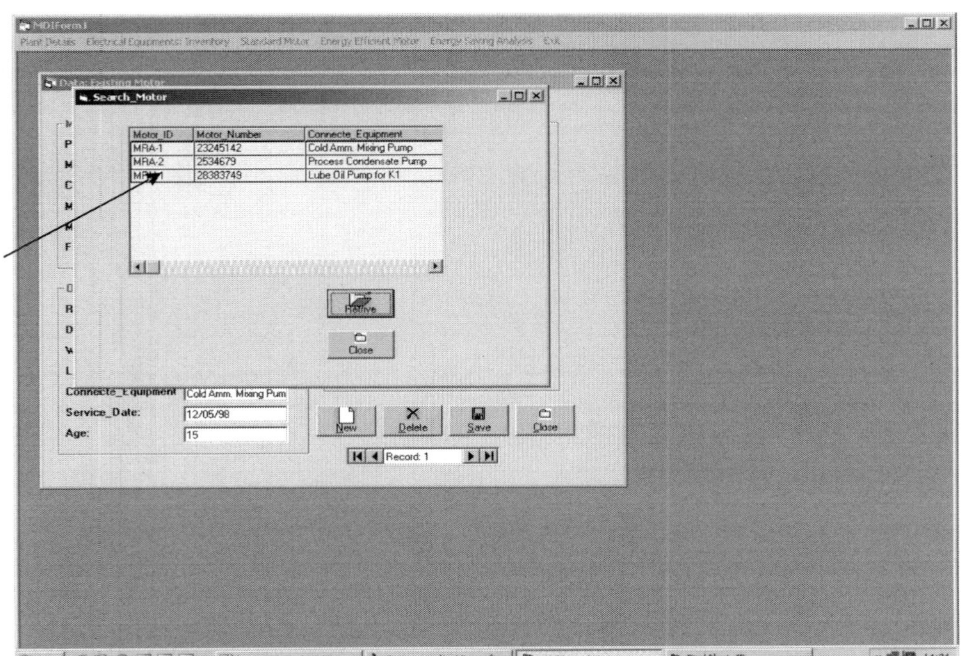

Fig. 2.13 Edit screen of motor

This screen shows the Motor-ID, motor number and the name of the connected equipment. Click on the "Motor-ID", for which editing is required. The arrow in Fig. 2.12 is pointing towards Motor-ID of last record. Clicking on that Motor-ID will result in transfer of all the records of that particular motor on main motor screen and the edit screen of motor will close automatically. The changes in the record as required can be incorporated on main motor screen, since all the records of that particular motor has already been transferred on main motor screen. Click on the "Save" button after making the changes to save the edited record in the data bank.

- Third entry is code. It is a normal practice in many industries to codify all the installed equipments.
- Most of the other parameters required to be filled in the other boxes can be picked either from name plate or from test certificate of the motor. The other data required to be filled are general data like running hours in a year, duty cycle, winding condition i.e. motor is re-winded one or still has its original windings, service date i.e. date on which motor is serviced and age of motor i.e. number of years of operation of motor. Two important parameters are loading of motor and efficiency of motor at that particular load. To get these

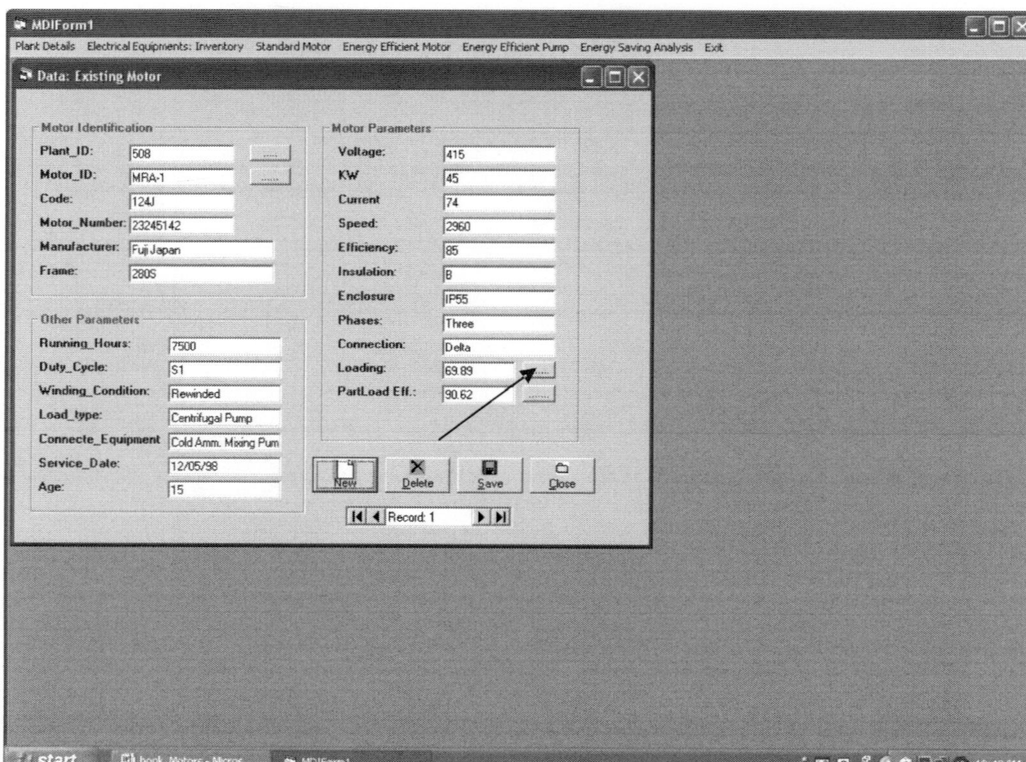

Fig. 2.14 Main motor screen with arrow indicating loading calculation box

data, field measurement of certain parameters of motor at load as well as at no load condition is required. The data can also be taken from the test certificate or load vs efficiency curve provided by the manufacturer, but if the motor is old and re-winded one, then to have accurate analysis, it would be better to measure the loading and efficiency parameters. To determine the loading of the motor, click on the button provided at the right hand side of the box titled "Loading" as shown in Fig. 2.14 by an arrow.

- A screen as shown in Fig. 2.15 opens up.

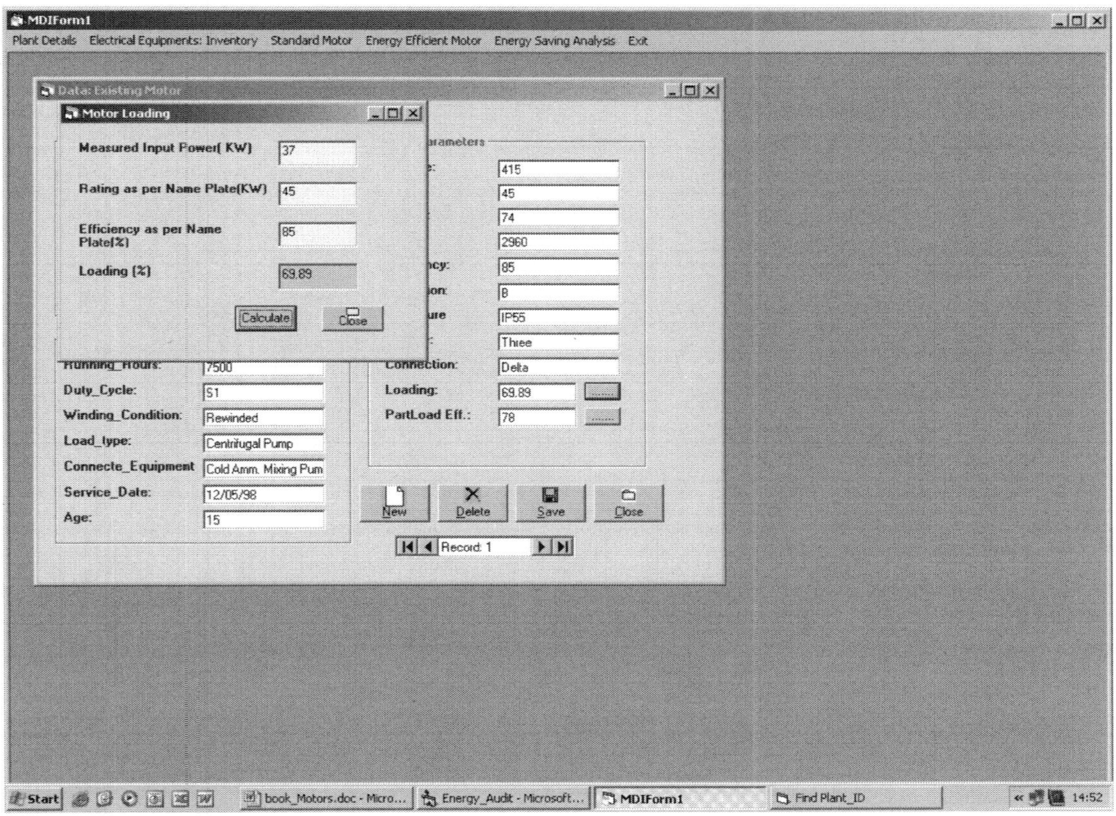

Fig. 2.15 Calculation of loading of motor

Measure the input power (KW) of the motor. KW rating and efficiency of the motor are obtained from the name plate or from test certificate. Enter these values in the respective boxes and click on the "Calculate" button. The loading of the motor will be calculated. This value will be copied in the empty box titled "Loading" on the main motor screen.

Example 2.1

Calculation of loading of the motor.

Input (measured) power = 37KW

Rated power = 45KW

Efficiency at rated load = 85%

$$\% \text{ Loading of motor} = \frac{\text{Input power at existing load} \times 100}{\text{Rated KW/Efficiency at rated KW}}$$

$$= \frac{37 \times 100}{45/0.85} = 69.89\%$$

Now, let's calculate efficiency of the motor at its operating load.

- Click on the button provided near the right hand side and a load efficiency box will appear as shown in Fig. 2.16 by an arrow.

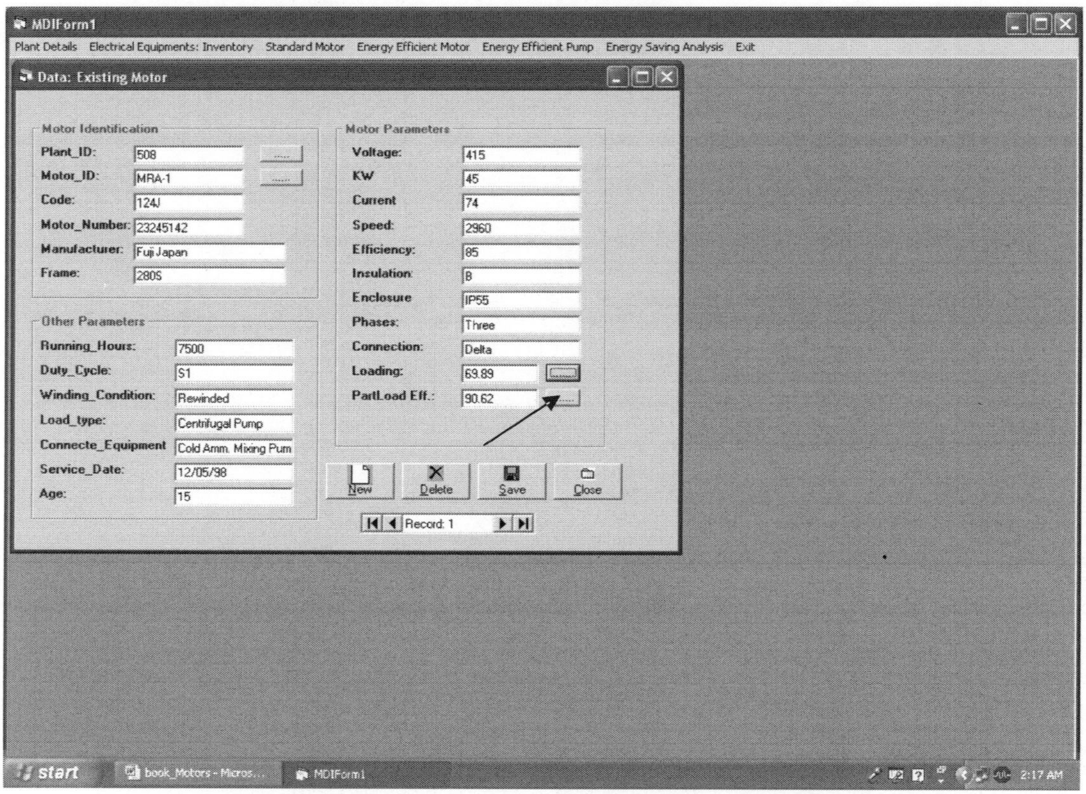

Fig. 2.16 Main motor screen with arrow indicating part load efficiency calculation box

- A screen as shown in Fig. 2.17 opens up. Various parameters like winding resistance, winding temperature, no load current, load current etc. are to be measured and put in the specified boxes on the screen.

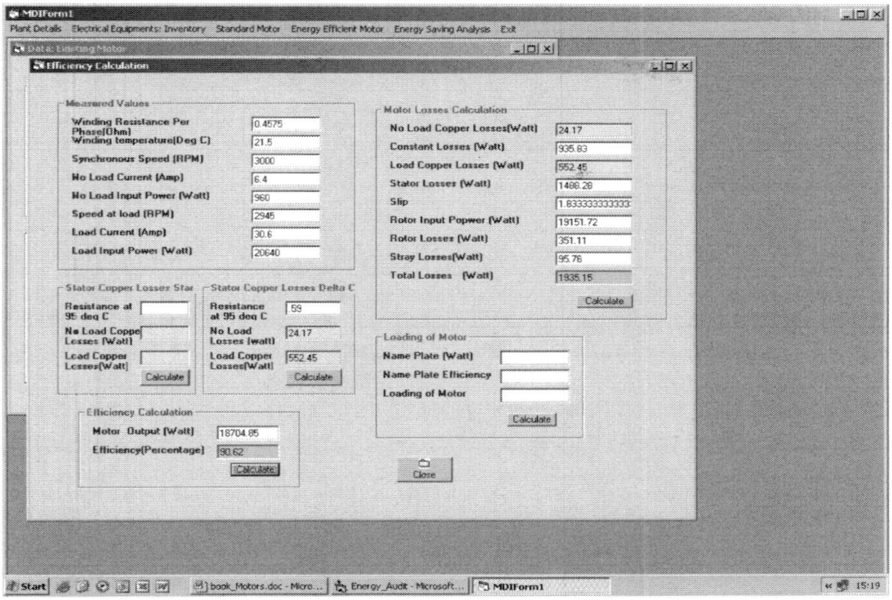

Fig. 2.17 Calculation of part load efficiency of motor

If the winding is connected in star, no load as well as load copper losses can be calculated by clicking the "Calculate" button provided in the frame titled as "Stator copper losses: Star Connected".

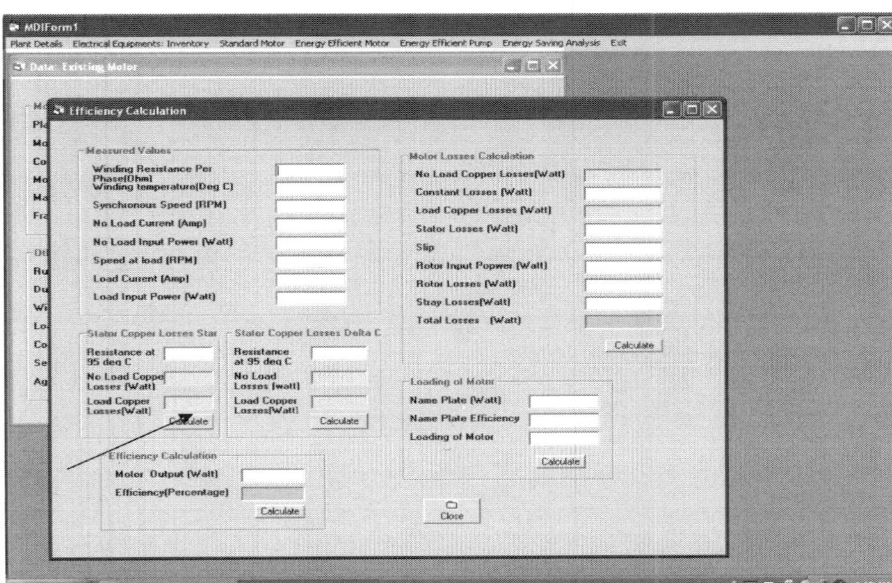

Fig. 2.18 Motor efficiency calculation screen with arrow showing calculation button when motor winding is connected in star

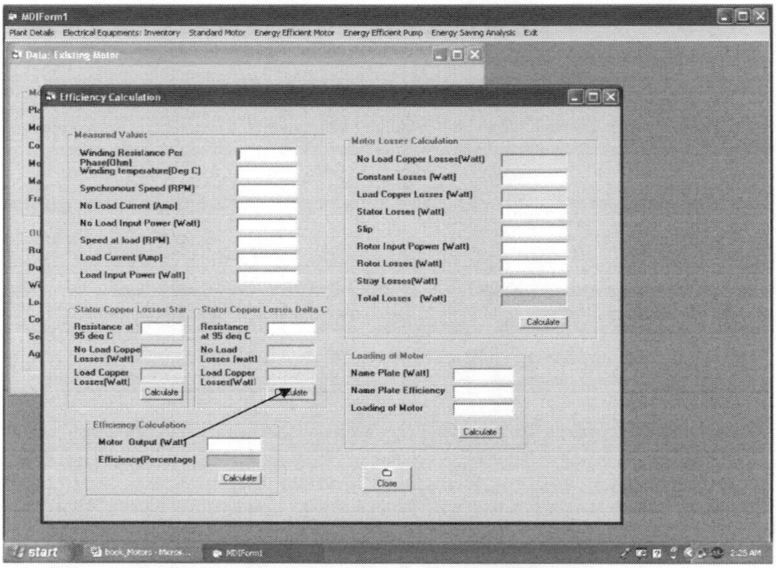

Fig. 2.19 Motor efficiency calculation screen with arrow showing calculation button when motor winding is connected in delta

Similarly, for delta connected winding, the copper losses are calculated by clicking on "Calculate" button provided in the frame titled as "Stator copper losses: Delta Connected". Refer Fig. 2.19. The values are automatically copied in the blank boxes titled as "No load and load copper losses" as shown in Fig. 2.20 by a solid and dotted arrow.

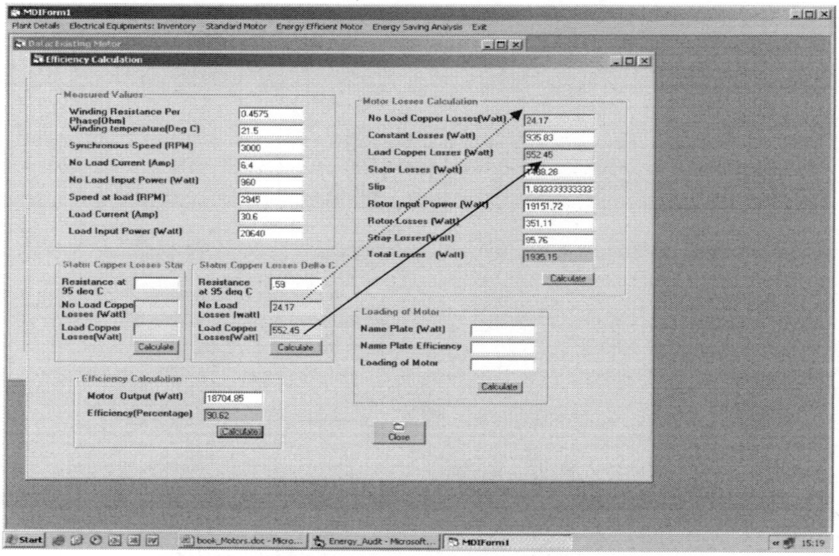

Fig. 2.20 No load and load copper losses

- Click on "Calculate" button provided in motor losses calculation frame to calculate the total losses. After that click on "Calculate" button provided in efficiency calculation frame to calculate the efficiency of the motor. The efficiency of the motor will be calculated and the value will be copied to main motor screen.

Example 2.2

Calculation of efficiency of a 2 pole 18.5KW motor.

Measured winding resistance at 21.5°C = 0.305 ohm (between terminals)
Motor is delta connected, therefore winding resistance per phase
$= 3 \times 0.305/2$
$= 0.4575$ ohm
Winding resistance at 95°C

$$= \frac{(235 + 95) \text{ x } 0.4575}{(235 + 21.5)} = 0.59 \text{ ohm}$$

Measured no load current = 6.4 amp
Measured no load input = 960 watt
No load stator copper losses $= (6.42)^2 \times 0.59 = 24.16$ watt
Constant losses $= 960 - 24.16 = 935.84$ watt
Measured value of input load current = 30.6 amp
Stator copper losses $= (30.62)^2 \times 0.59 = 552.45$ watt
Measured input power = 20640 watt
Rotor input $= 20640 - (935.84 + 552.45) = 19151.7$ watt
Measured slip = 1.8333
Rotor losses $= .018333 \times 19151.7 = 351.1$ watt
Stray load losses = 0.5% of rotor input
$= 0.005 \times 19151.7 = 95.76$ watt
Total losses = Stator constant losses (Core losses) + Stator copper losses
+ Rotor losses + Stray losses
$= 935.85 + 552.45 + 351.1 + 95.76$
$= 1935.16$ watt
Motor output = Input power – Total losses
$= 20640 - 1935.16 = 18704.84$ watt
% Efficiency = 100 × Motor output/Motor input
$= 18704.84 \times 100/20640$
$= 90.62\%$

The value of calculated efficiency is same as calculated by software and the same also appears in Fig. 2.16.

- The next step is to develop data bank for standard efficiency motor.

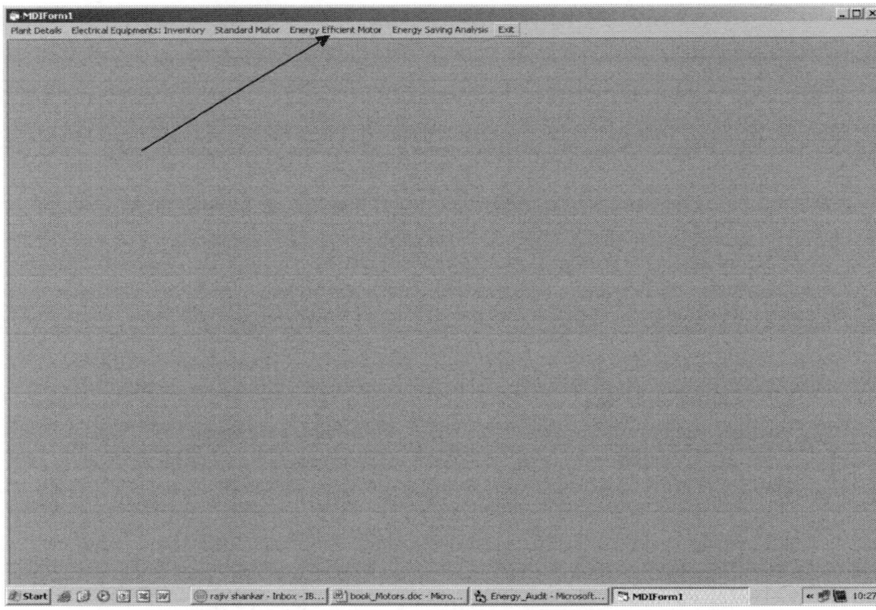

Fig. 2.21 Arrow showing menu for standard efficiency motor on main screen

- Click on "Standard Motor" of the main menu, as shown by an arrow in Fig. 2.21 and a screen will appear as shown in Fig. 2.22.

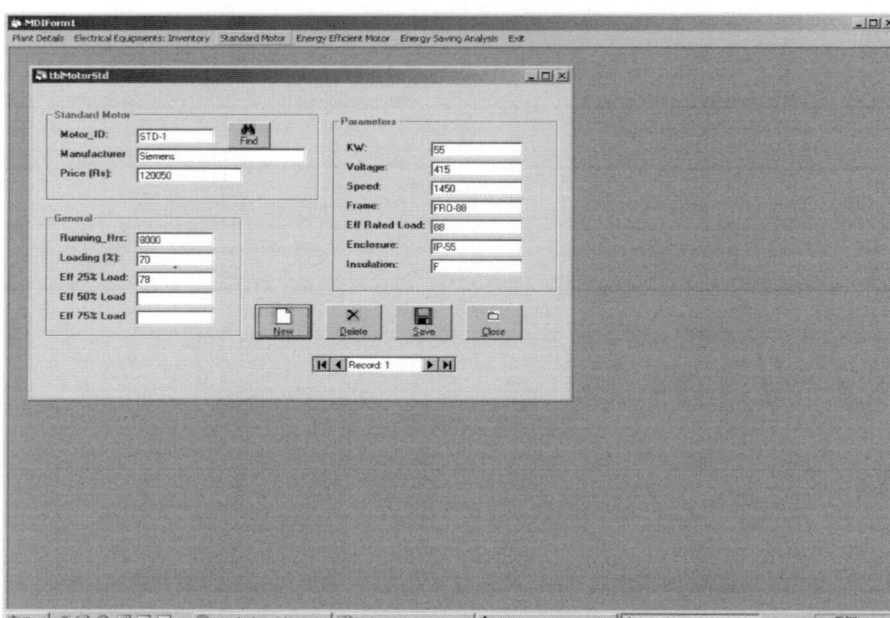

Fig. 2.22 Data entry screen for a standard efficiency motor

The procedure of entry of information and storing is similar as mentioned earlier. The information required to be stored can be obtained either from name plate of the motor or from test certificates of the motor supplied by the manufacturer. In a similar way, data bank for high efficiency motor can also be created. Refer Fig. 2.23 and Fig. 2.24.

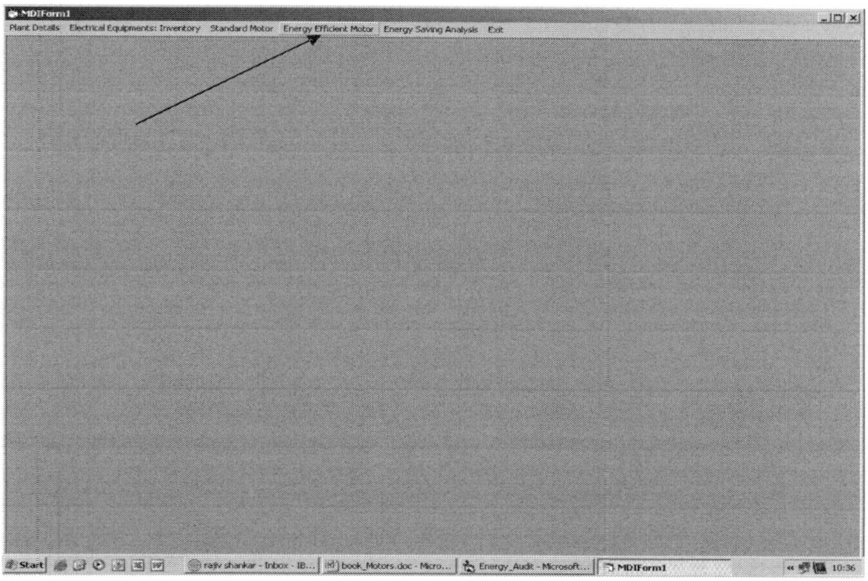

Fig. 2.23 Arrow showing menu for high efficiency motor on main screen

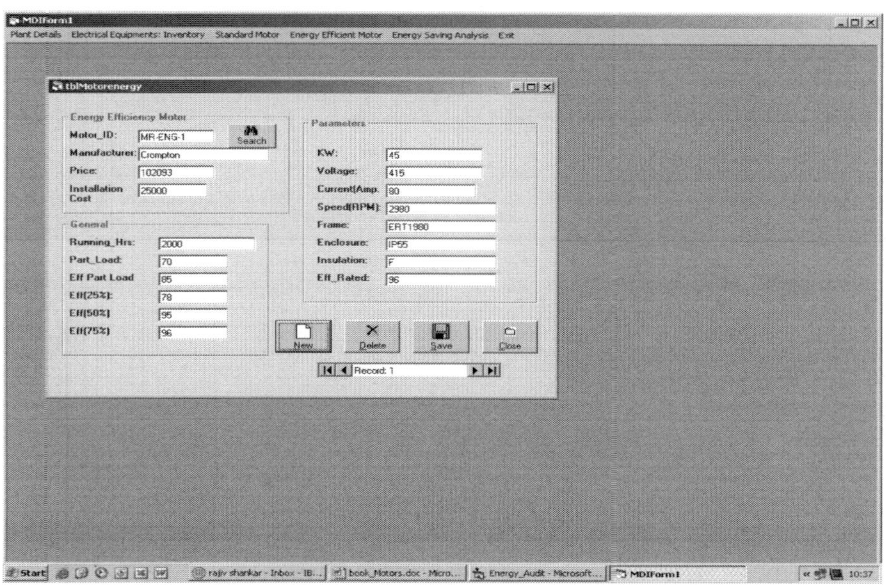

Fig. 2.24 Data entry screen for high efficiency motor

- Now, we have a data bank for three types of motors i.e. existing motor, new standard efficiency motor and new high efficiency motor. These data banks will be helpful in identifying the inefficient motors and then making an analysis for saving in energy, if the identified inefficient motor is replaced with the energy efficient motor/standard efficient motor. The software also calculates the payback period for the investment made.
- Let's go to the analytical part of the software. Click on "Energy Saving Analysis" of main menu, a submenu will appear, as shown in Fig. 2.25. Click on "Energy Efficient Motor" of the submenu as shown by an arrow.

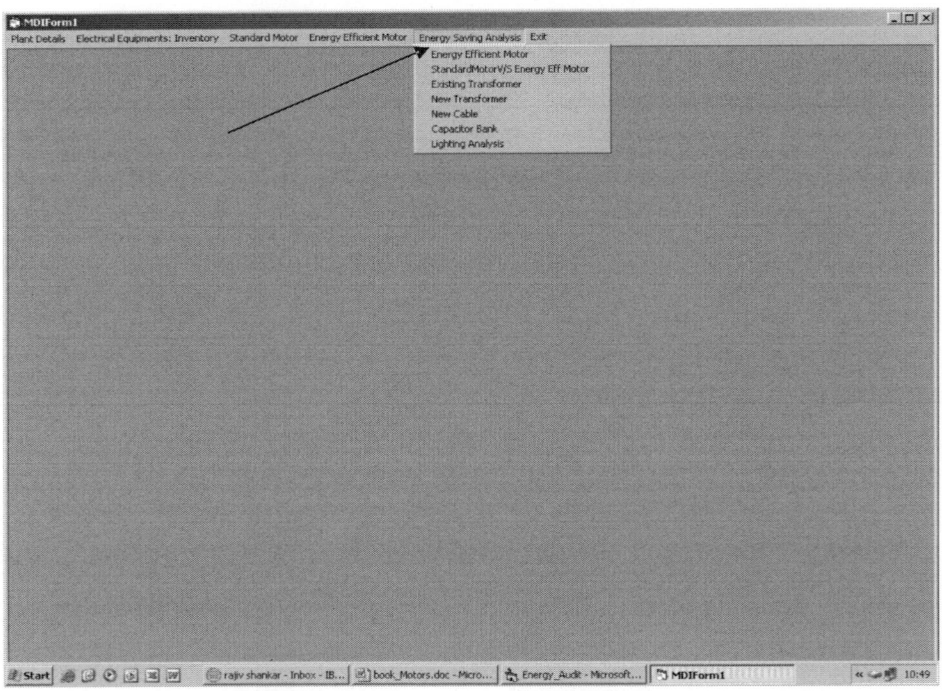

Fig. 2.25 Arrow showing submenu for analysis of motor

It will open a screen as shown in Fig. 2.26. Let's have a look on the contents of the screen. We find a box and a button as marked by a solid line arrow. After clicking of button another screen will open, which will have Plant-ID and name of all the plants which are in the database. Click on the Plant-ID and the name of the plant will be automatically copied in the respective box. The name of the selected plant will appear in the box provided on the right hand side of the combo-box. The selected plant should be of that plant from where motor will be taken for analysis. To get the data of the motor for which analysis is required, click on the "Data bank" button as shown by dotted line arrow in Fig. 2.26.

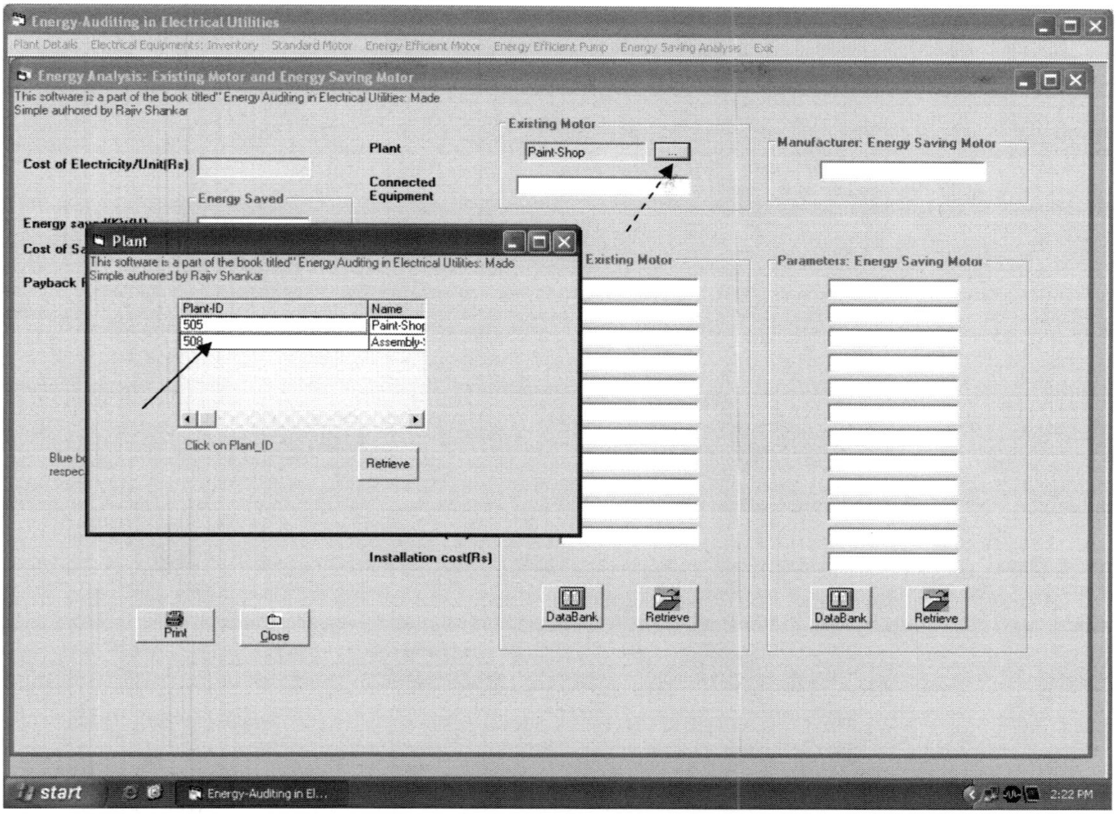

Fig. 2.26 Motor analysis screen

The clicking action will open a screen as shown in Fig. 2.27. This screen is for retrieving of details of motor as per the specified criteria. This type of data retrieval is called “Retrieval of data by generation of query”. To generate a query for retrieval of motor details, following steps are to be taken.

- Select the plant by clicking the button as shown in Fig. 2.27 by a dotted arrow and as per procedure which is mentioned earlier.
- The respective boxes are required to be filled with the range of voltages. For example, if only LT motor are required to be selected, both field i.e. “From and To” shall be filled with LT voltage. For example, for an LT motor range From : 415 To : 415 or From : 440 To : 440 as the case may be, is to be filled in the respective boxes. If we want to select LT as well as HT motors, then the range From : 415 To : 3300 or any other range may be filled in the boxes. Similarly, the range for KW and efficiency are also to be filled.

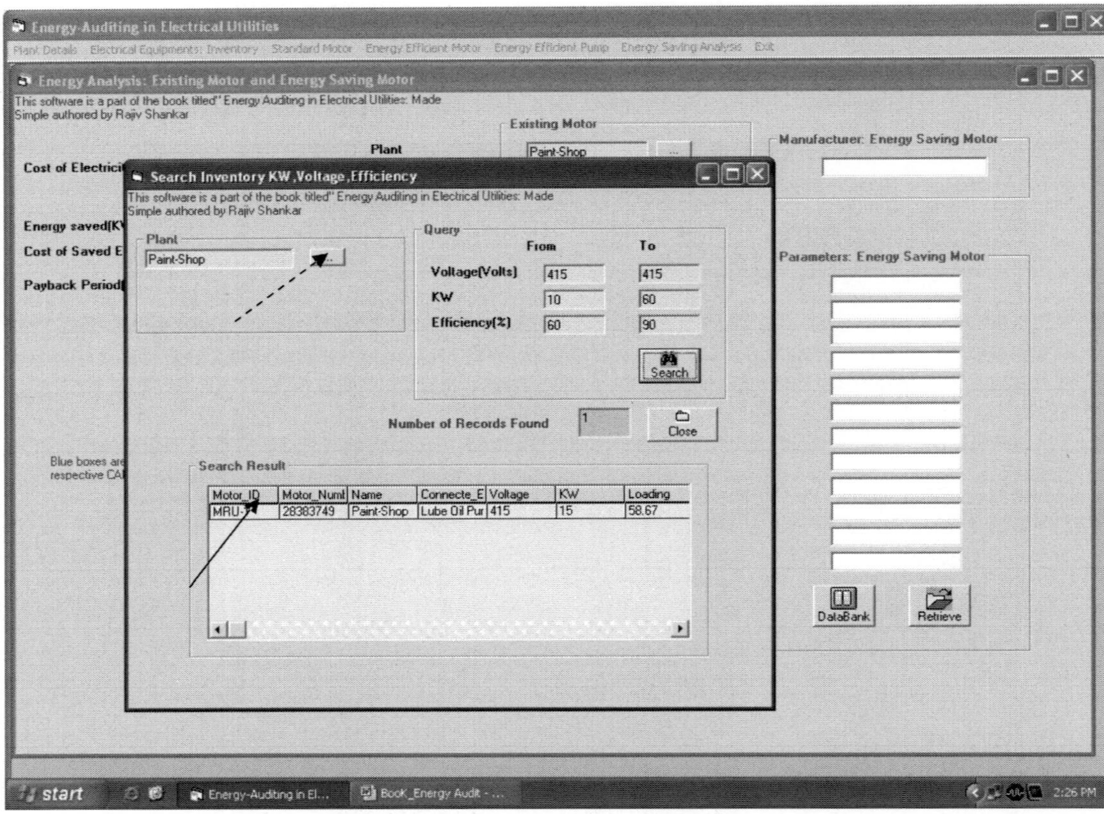

Fig. 2.27 Query generation screen

- After fixing the search criteria, click on "Search" button, as shown in Fig. 2.27 by a dotted arrow. The blank grid will be filled with the motors details, which are available in the data bank and whose parameters fall within the range as fixed in search criteria. In short first fix the search criteria and then based on that criteria motors are selected. Whenever, energy auditing is done, first the motors of large rating and of less efficiency are selected. The parameters of the selected motor needs to be transferred to the analysis screen. To accomplish this, click on the "Motor-ID" for which analysis is to be done as shown by solid arrow in Fig. 2.27.On clicking, all the parameters of the selected motors will be transferred to the motor analysis screen, however, to retrieve these parameters, click on the "Retrieve" button of the analysis screen as shown by dotted arrow in Fig. 2.28. On clicking the retrieve button all the blank boxes will be filled with the parameters of the selected motor.

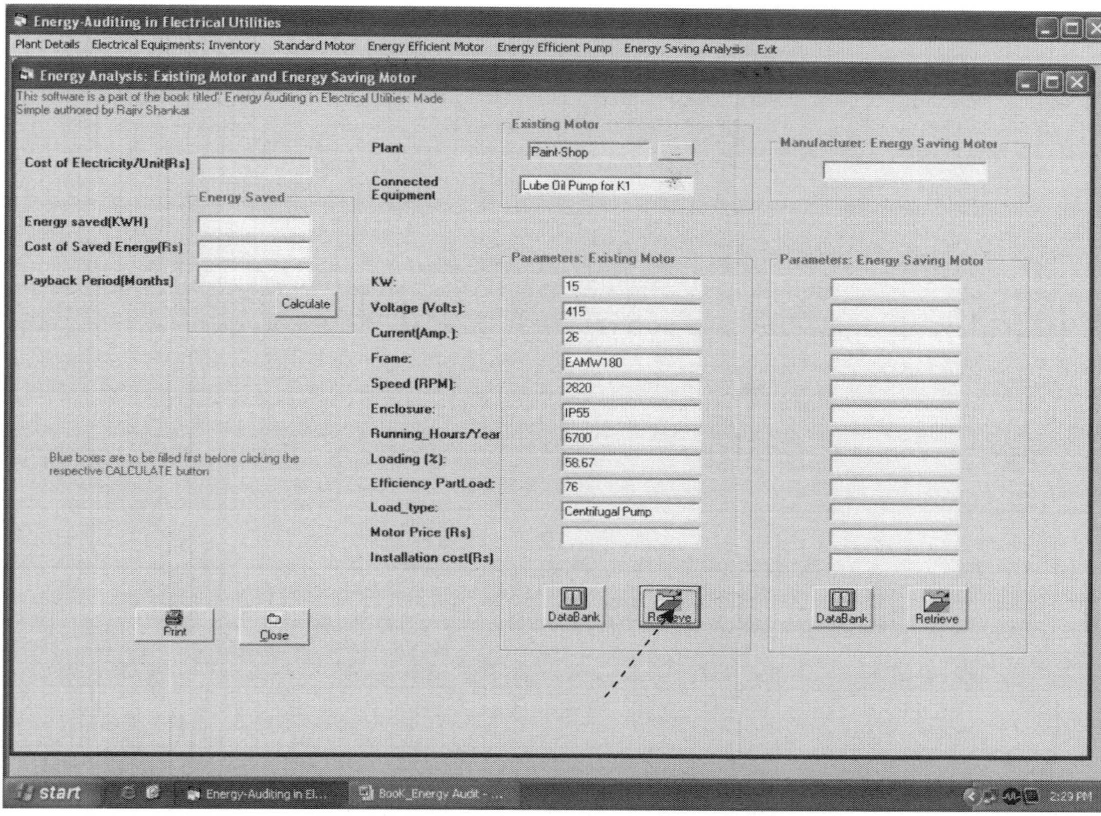

Fig. 2.28 Retrieval of parameters of selected motor

- In the next step a search for a similar rating motor is made from the data bank of energy efficient motor so that both the motors can be compared as far as energy consumption is concerned. The procedure for retrieving parameters of a similar rating of energy efficient motor or standard efficiency motor is same as mentioned earlier.

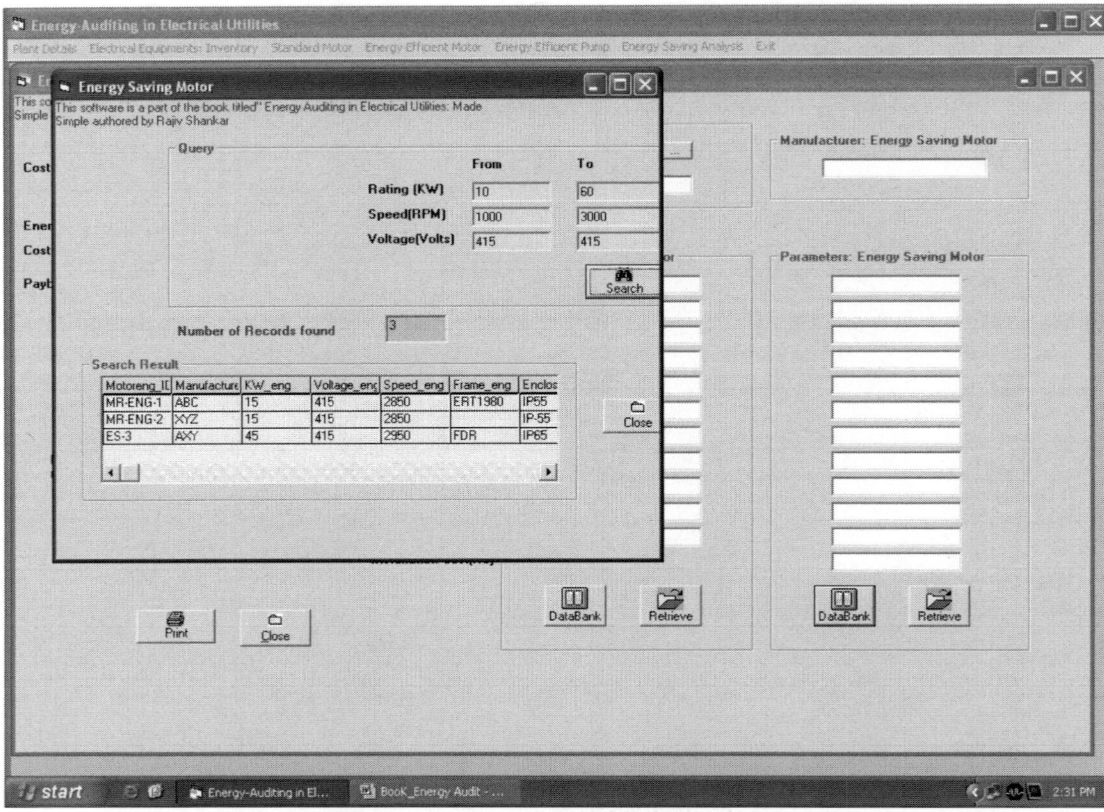

Fig. 2.29 Screen for searching energy efficient motor

After retrieval and transferring of parameters of both the motors to motor analysis screen, we can proceed for analysis. Refer Fig. 2.30.

- Fill the average Kilowatt hour (unit) charges of electricity in the specified box, as marked by an arrow in Fig. 2.30 and click on the "Calculate" button. The energy saved, cost of saved energy as well as simple payback period is calculated. The result can be printed by clicking the "Print" button as shown in Fig. 2.31.

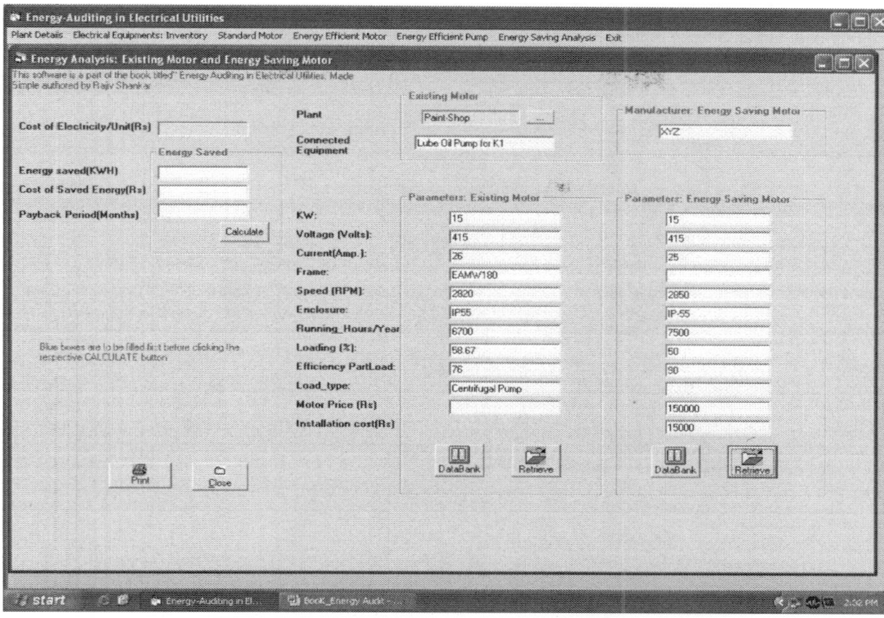

Fig. 2.30 Calculation of energy saving and payback period

The above analysis will help in decision-making for replacement of existing motor with a standard efficiency motor or a with energy efficiency motor.

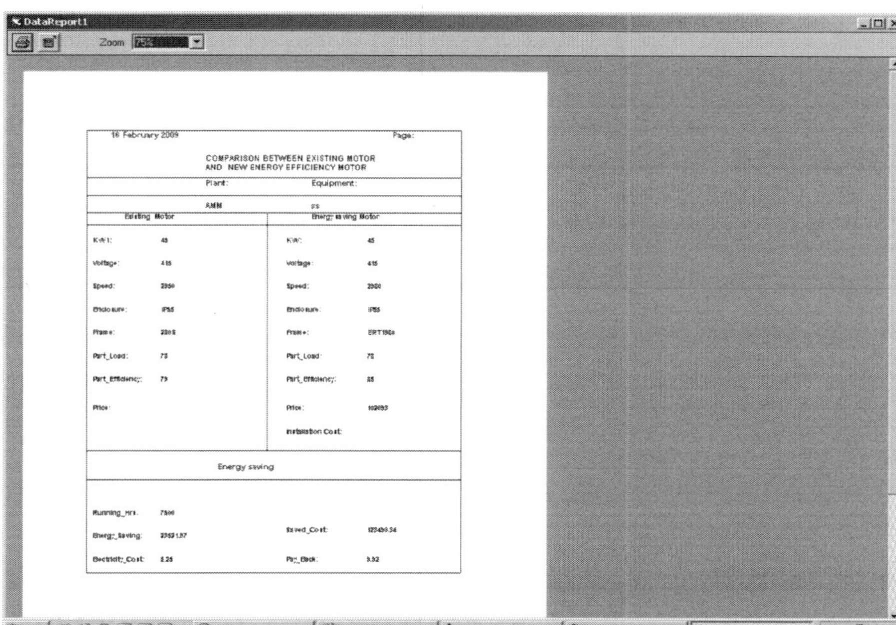

Fig. 2.31 Printing of motor analysis result

Let's take an example.

Example 2.3

Motor rating = 45KW
Efficiency of existing motor = 79%
Efficiency of the energy efficient motor = 96%
Running hours = 7500 hours
Electricity charges = Rs5.25
Loading of motor = 78%
Cost of energy efficient motor = Rs230000.00
Energy saved = 0.78 × 7500 × (45/0.79 – 45/0.85)
= 23521.81KWH
Cost of energy saved = Rs5.25 × 23521.81
= Rs123489.54
Payback period = 230000 × 12/123489.54
= 22.35 months

The same result is obtained by the software as shown in Fig. 2.30

- Similar analysis can be done for selection of a standard and an energy efficient motor for new installation. Click on the "Energy Saving Analysis" of the main menu. A submenu will appear, click on "Standard Motor/Energy Efficiency Motor" refer Fig. 2.32. On clicking, a screen will appear as shown in Fig. 2.33. Rest of the procedure like generation of query, retrieving the query results and analysing the result is same as mentioned earlier. Refer Fig. 2.33 to Fig. 2.35.

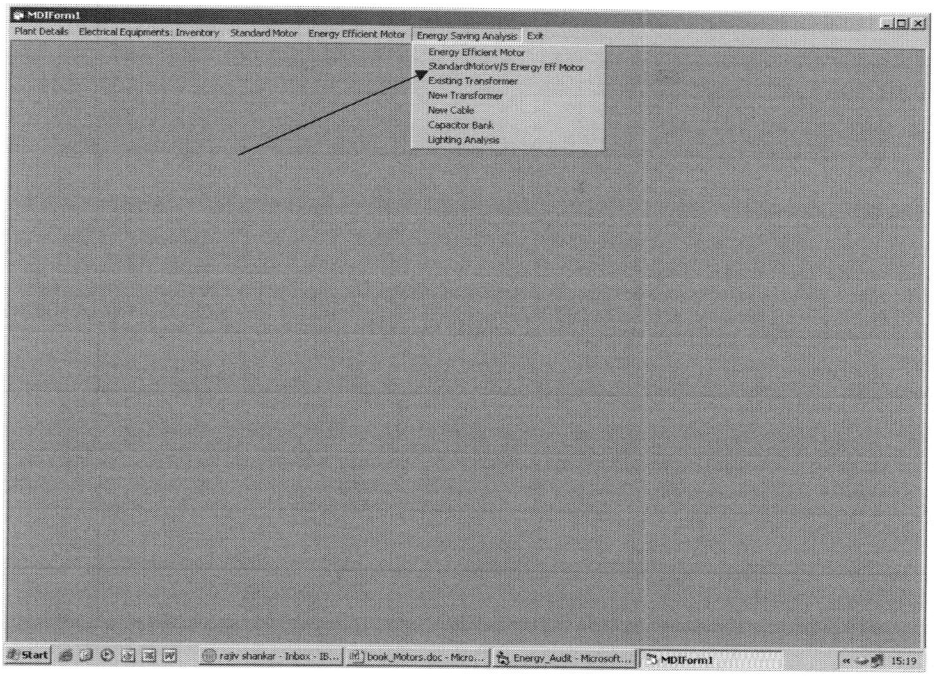

Fig. 2.32 Comparison of energy saving motor vs standard motor

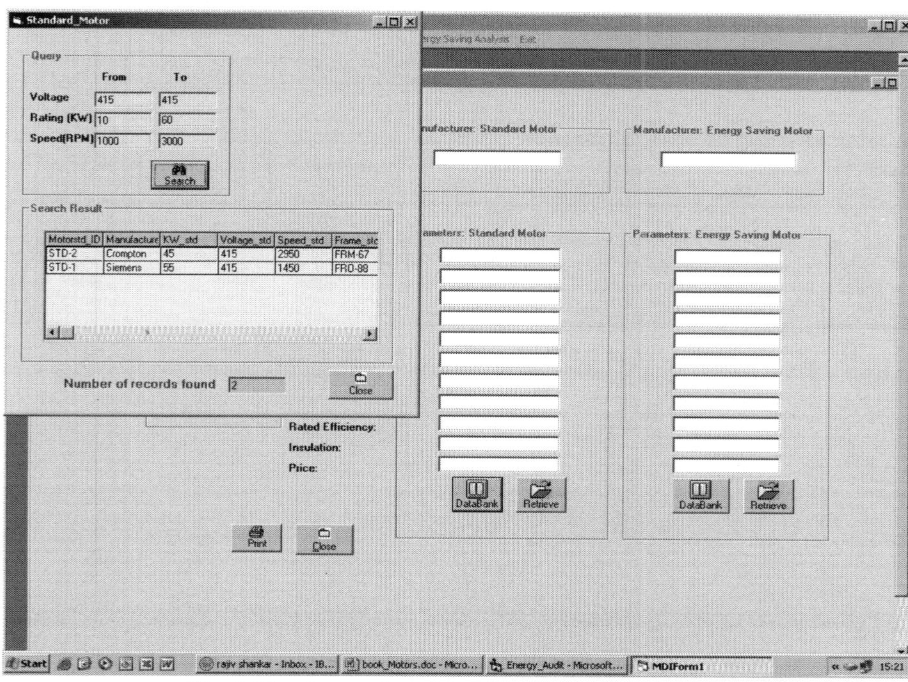

Fig. 2.33 Generation of query for standard motor

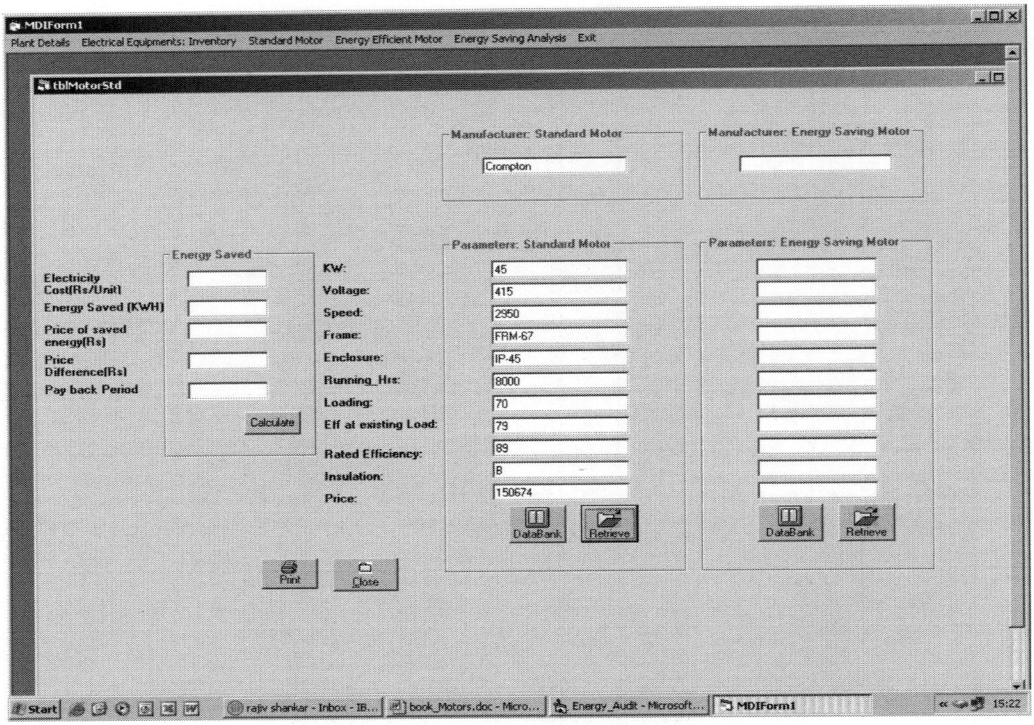

Fig. 2.34 Retrieval of data for standard motor on main screen

The next important step is to print the results of all the analysis so that it can be attached with the energy audit report. The same can be done by clicking on "Print" button provided on each motor analysis screen. The printed results is shown in Fig. 2.36.

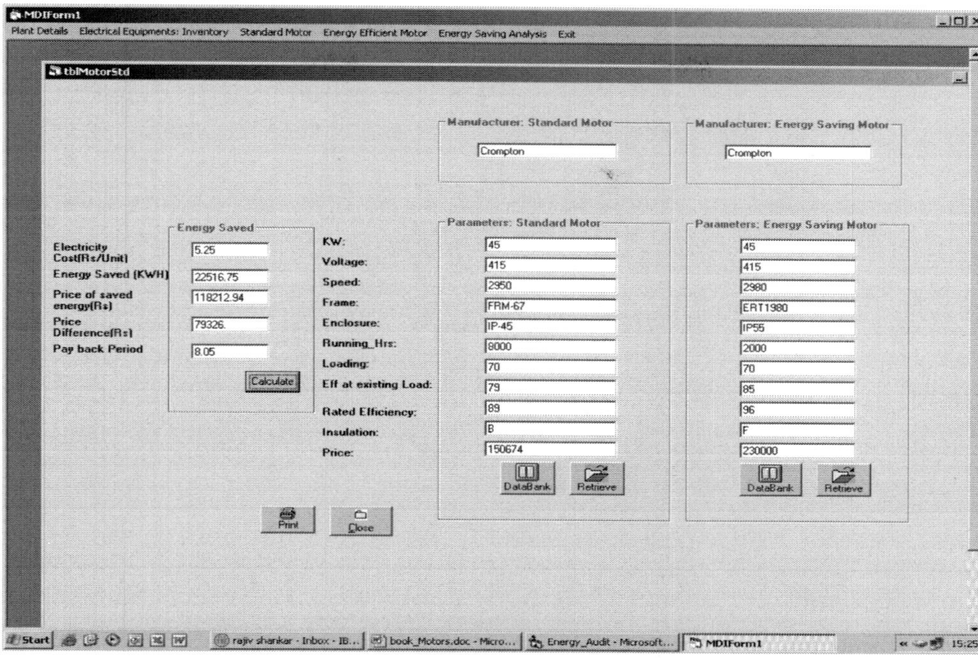

Fig. 2.35 Comparison between standard motor and energy efficient motor

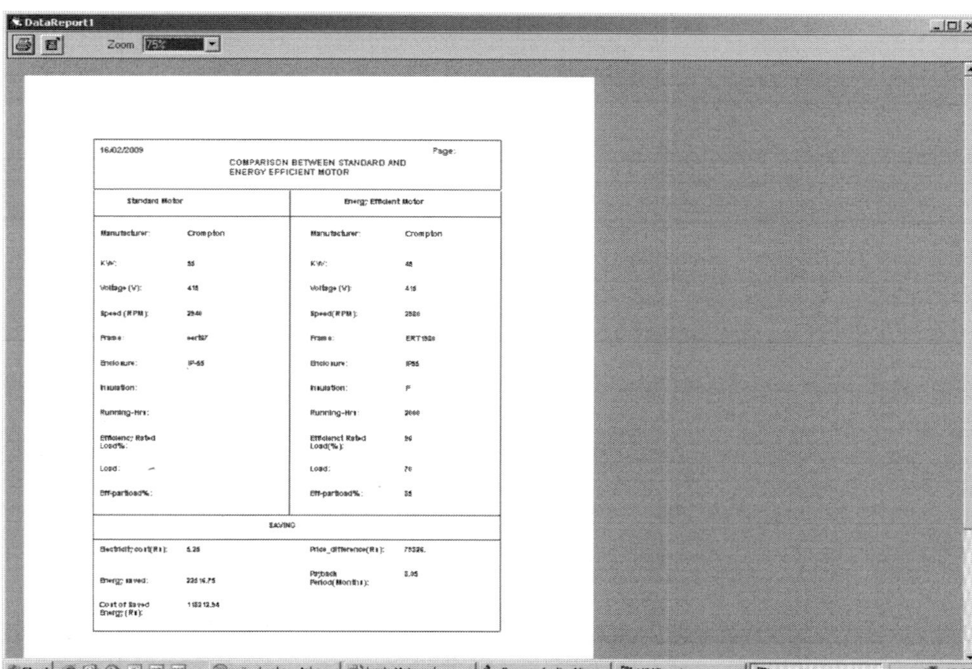

Fig. 2.36 Printing obtained from comparison of standard motor and energy efficient motor

CHAPTER

3

Transformer

3.0 Introduction

Transformer is an integral part of an electrical sub-station. Bulk of industrial as well as commercial load is supplied at medium or high voltages as per the norms of the state electricity boards, because it is always advantageous to supply bulk load at medium or high voltage due to many reasons like reduction in transmission losses, better voltage regulations etc. Step up transformers are required to step up the voltage at required level for transmission of power. However, the consumer equipments are not suitable for the voltage which is received at their premises and it has to be stepped down. The distribution of transformer is required to step down the voltage up to a level which is suitable for the equipments installed at the consumer premises.

In this way every electrical sub-station where voltage needs to be stepped up or stepped down, transformer is provided. It may have a single transformer or many transformers, as per the requirement of the electrical distribution system. The medium voltages at which electricity is transmitted are 66KV, 33KV, 22KV and 11KV. These voltages

Fig. 3.0 Transformer

are stepped down to 6.6KV and 3.3KV as per the requirement at consumer end. The low voltage system has 415V.

The transformer is a very efficient piece of equipment and its efficiency varies between 97% to 98.8% depending upon the size of transformer, loading and power factor of the system.

In this chapter we will be mainly discussing about the losses and efficiency of the transformer, ways and means to decrease the losses and increase in efficiency level, evaluation of the ownership cost of the transformer while purchasing a new transformer. We will also go through a little bit of basic principle of working of transformer, ways of energy conservation in transformers in an electrical system and finally we will go through the various steps for calculation of energy saving in a distribution transformer with the help of software included with this book.

A person without background of electrical engineering will also find it easy to understand the basic energy conservation aspects related with the transformer. Therefore, no complex formula is derived and mathematics involved is very simple.

3.1 Basic Principle of Working of a Transformer

Whenever, a voltage is applied to a coil, current flows through it and produces a magnetic field. The strength and direction of magnetic field depends on the strength and direction of the current flowing through the coil. If the current is a pulsating current i.e. AC (Alternating Current) the magnitude as well as direction of the magnetic field also keeps on changing. If a coil is placed in that magnetic field, an emf (voltage) will be induced in the coil. The magnitude of the emf will depend on the linkage of the flux (magnetic field) and its magnitude.

In actual practice, half of the primary and half of the secondary winding is around each limb of the core.

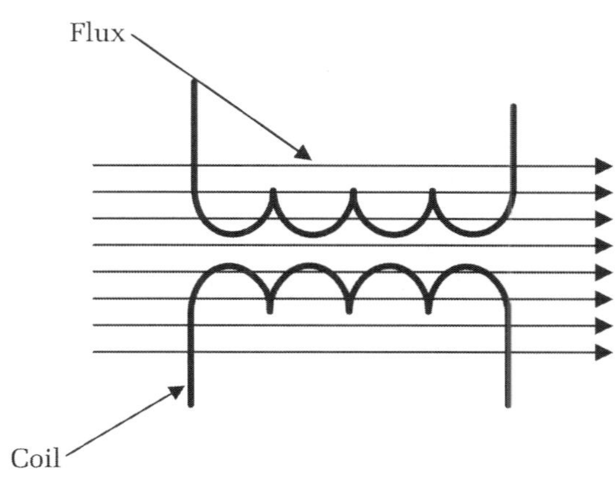

Fig. 3.1

Sufficient insulation is provided between the core and between primary and secondary winding. The core is made of laminated strips of silicon steel material and the windings are of copper. The construction of three phase transformer is slightly different from the single phase transformer. Three phase transformer consists of three limbs in the core instead of two. The reason of such type of construction is to avoid leakage of flux and to have maximum linkages between the two coils.

The voltage developed by transformer action is given by the following formula:

$V = 4.44fNB_m$
f = Frequency
N = Number of turns
B_m = Flux density in the core, telsa (weber/m^2)

From the above formula, we find that normally the frequency and the magnetic field remain constant and the voltage induced depends on the number of turns in the coil. If number of turns are more, voltage induced will be more and vice versa.

Due to this reason, high voltage winding has more number of turns than the low voltage winding.

3.2 Transformer Losses

The transformer losses can be classified as fixed losses and variable losses.

3.2.1 Fixed Losses

Fixed losses do not change with variation of transformer load and remain constant even though the transformer is not connected to load. Fixed losses consist of hystersis losses and eddy current losses. Dielectric losses are also fixed losses but since, it is very less, it can be ignored.

Hystersis Losses

As mentioned earlier that the magnetic field in the transformer core is of pulsating nature, magnetization and demagnetization of the core take place. Due to this magnetization and demagnetization the orientation of molecular magnetic domain keeps on changing. It has to realign itself when the magnetic force changes. This alignment and realignment of magnetic domain causes losses, which is called hystersis losses. The hystersis loss in the core depends on the type of core material. Hystersis losses can be expressed as:

$H = K_h f B_m^x$
H = Hystersis loss in watts/Kg
f = Frequency, Hz

Kh = Constant which depends on the molecular characteristic of the material and also on the nature of the magnetic field i.e. pulsating or rotary in nature.

B_m is the flux density and the value of x lies between 1.5 and 2.

Eddy Current Losses

Eddy currents are generated locally in the core due to pulsation in the magnetic field and the losses can be expressed as:

$E = Ke\, t^2 f^2 B_m^2 / r_o$

Ke = Constant

t = Thickness of the material (mm)

f = Frequency, Hz

B_m = Density of the magnetic field (weber/m²)

r_o = Resistivity of the core

As seen in the above formula, eddy current losses can be reduced by having a thinner and high-resistivity material.

The estimation of core losses is normally done by the graphs, which are drawn by doing experiments on a specimen in the laboratory.

3.2.2 Load Losses

Load losses are those losses which take place due to current flowing in the primary and secondary winding. It may also be called the Copper Losses and depends on the resistance of the primary and secondary winding of the transformer and the current flowing through these windings. The load losses vary as per loading of the transformer.

The value of the resistance depends on the cross section area of the conductor and the length of the conductor. The formula for calculating resistance is given below:

$R = r_o\, l/a$ (ohm)

l = Length of the conductor (metre)

r_o = Specific resistance

a = Cross section area of the conductor (metre²)

The resistance also depends on the temperature and increases in the temperature. The following formula can be used for calculation of resistance at a specified temperature.

$R_1 = R_0\,[T_1+235]/[T_0+235]$

R_1 = Resistance to be calculated at temperature T_1

R_0 = Resistance at temperature T_0

The copper losses can be calculated by the following formula:

Copper losses = $i^2 r$

Where i = Current flowing through the winding and r is the resistance at the specified temperature.

3.3 Evaluation of Transformer Losses

The life of a transformer is 25 to 30 years or more, therefore while purchasing a transformer the cost of losses over its life period should also be evaluated. The combine cost i.e. initial cost of the transformer which is incurred at the time of purchase of the transformer and the cost of losses over its life period is called Total Ownership Cost (TOC). It is necessary to work out total ownership cost of the transformer while purchasing a transformer. To treat the cost of losses of the transformer over its life span, present value method is adopted. In this method, present value of the cost of losses incurred during life span of the transformer is calculated with the help of following formula:

Present value of the cost of losses of the transformer

$= ((1 + i)^n - 1)/i(1 + i)^n$ ....(A)

i = Rate of discount

n = Number of years for which transformer will remain in operation.

For example, with the inflation rate of 10%, the value of 1000 rupees after one year will be:

$= 1/(1 + i)^n$

$= 1/(1 + 0.1)^1$

$= 1/1.1 = 0.909 \times 1000$

= Rs909

If we look it another way, Rs909 deposited in bank at interest rate of 10%, it will be Rs1000 after one year. Similar calculation can be done for 2nd year, 3rd year and so on.

While evaluating the TOC (Total Ownership Cost) of transformer, the present value of the losses of the transformer is calculated at certain discount rate over its life time with the formula (A).

Many times, it has been observed that though the initial cost for procurement of new transformer of one manufacturer is lower than the another transformer, but the TOC may be more due to more losses. Therefore, the transformer of less TOC should be selected. Let's take an example.

Example 3.1

We have two transformers, Transformer 1 and Transformer 2.

Transformer 1 has following details.

Transformer 1:

Purchase price = Rs3.00lac
No load losses = 1650 watt
Load losses = 9500 watt
The transformer is operating at 70% of its rated capacity.
Load losses at 70% load = $0.70^2 \times 9500$ = 4655 watt

Transformer 2 has following details:

Transformer 2:

Purchase price = Rs3.15lac
No load losses = 1450 watt
Load losses = 7500 watt
The transformer is operating at 70% of its rated capacity.
Load losses at 70% load = $0.70^2 \times 7500$ = 3675 watt
The electricity cost is Rs5.25.
Discount rate =10%
Life of both the transformers is 15 years and the transformer remains on load for 8500 hrs in a year.

Calculation

Transformer 1:

Cost of no load losses per year = $1.650 \times 5.25 \times 8500$ = Rs73631
Present value of these losses = $((1 + 0.1)^{15} - 1))/((0.1)(1 + 0.1))^{15}$ = 7.606
PV cost of no load losses = 7.606×73671= Rs560039
Similarly, we can calculate the PV of cost of load losses
= $7.606 \times 4.655 \times 8500 \times 5.25$
= Rs1579990.00
TOC = 300000 + 560039 + 1579990
= **Rs2440029.00**

Transformer 2:

Cost of no load losses per year = $1.45 \times 5.25 \times 8500$ = Rs64706
Present value of these losses = $((1 + 0.1)^{15} - 1))/((1 + 0.1)^{15}(0.1))$ = 7.606
PV cost of no load losses = 7.606×64706 = Rs492154.00
PV cost of load losses = $7.606 \times 3.675 \times 8500 \times 5.25$ = Rs1247360.00
TOC = 315000 + 492145 + 1247360
= **Rs2054514.00**

From the above example, it is clear that though the initial purchasing price of Transformer 1 is less than the Transformer 2, the total ownership cost of Transformer 2 is less than the Transformer 1, reason being both losses i.e. fixed as well as load losses in Transformer 2 is less in comparison with Transformer 1.

The above mentioned procedure is simple and can be used for comparison of two transformers. For simplicity, many assumptions are made like constant electricity cost, discount rate, loading of transformer etc. which may not remain constant over the life of the transformer. The temperature effect on the transformer is also ignored.

3.4 Reduction in Transformer Losses

Transformer losses can be reduced during design and construction stages itself. Core losses can be reduce by using low silicon steel as core material. Amorphous steel is also used to reduce the core losses. Further, improvement in cutting of laminations and assembly, use of thinner laminations and stepped lapped joints instead of butting the laminations joints improves the core losses.

Electrolyte grade, copper winding are used in place of aluminium winding to reduce the copper losses, since copper has lower resistance per cross section area in comparison to aluminium.

As far as procurement of new transformer is concerned, it should be chosen in such a way that the efficiency shall remain high at the load at which it is required to be operated. Normally, transformers operate at maximum efficiency at 40 to 50 percentage of load.

At the stage of designing of electrical system, it is decided that whether system will have a single or two or more than two incoming feeders. Accordingly, KVA rating of transformers also vary. Two feeders system is preferred, where load is of critical nature and continuity of power supply is required. In such a system, two incomers with two transformers are provided and each transformer is capable to take 100% of load in case of failure of any transformer. Load should be distributed between two transformers in such a way that equal in normal condition, each transformer shares almost equal load. Since each transformer shares equal load, it should have maximum efficiency at that load i.e. 50% of total load. Such type of system exits in process industries like fertilizer plants, chemical plants etc. and load remains constant throughout the period of loading of transformer.

Up till now, we have discussed about transformers which are yet to be procured and the electrical system is in design stage. Now, the question arises that how to improve the efficiency of a transformer which is already in service?

Energy auditors suggest various ways for reduction of losses and improvement in efficiency. Some of them are mentioned below:

- The load of lightly loaded transformer is distributed between other transformers and switching off that transformer so that the core losses become zero. The switching off of transformer is possible only when more than two transformers are provided.
- The load among transformers is redistributed so as to achieve load factor for each transformer, for which it has maximum efficiency so that it could operate either at its maximum efficiency point or near to it.
- As mentioned earlier also core losses depend on the primary voltage. Reduction in primary voltage also causes reduction in core losses, to some extent. But at the same time, the secondary voltage also reduces. Before, going for reduction in primary voltage, it is necessary to find out the type of load fed by the transformer. If the load is lighting load, then definitely reduction in energy consumption will take place on reduction of primary voltage. Voltage should not be reduced too much, otherwise flickering of lighting fixture may take place. On the other hand, if motors are connected and loaded up to 80% or more of its rated capacity, then reduction in voltage will cause increase in current which will increase the load losses. However, if the motors are lightly loaded then reduction in voltage will improve the power factor and efficiency of the motors.

3.5 Energy Conservation Building Code

These codes specify certain requirements for a transformer. Some of these requirements are mentioned below:

- Mandatory requirements
- Maximum allowable power transformer losses

The transformers must be selected in such a way that it has a minimum acceptable efficiency at the full load rating. The initial cost of the transformer and the present value of the cost of losses over its life span should be minimum. Transformer used in building shall be of high efficiency and constructed with high quality grain oriented, low loss silicon steel and virgin electrolytic grade copper. A certificate from the manufacturer should be obtained in this regard.

Table 3.1 indicates the maximum allowable losses of 11 and 22KV transformers.

Table 3.1 Maximum Allowable Losses of 11 and 22KV Transformers

Transformer Capacity KVA	*Max. Allowable Losses at Full Load in % of Rating 11KV*	*Max. Allowable Losses at Full Load in % of Rating 22KV*
100	2.5	2.7
160	2.3	2.2
250	2.1	1.8
400	1.5	1.5
630	1.4	1.5
800	1.4	1.5
1000	1.2	1.2

- Measurement and reporting of transformer losses

Measurement of losses shall be done with a calibrated digital meter of accuracy class 0.5 or better and certified by the manufacturer. All transformers of 500KVA and above shall be equipped with metering class current transformers and voltage transformer in additional to the requirements to the utilities to monitoring the periodic losses in the transformer.

3.6 Conclusion

Transformer is a static and efficient piece of electrical equipment. Generally transformers operate on maximum efficiency at 40% to 50% of load. Evaluation of losses during procurement of transformer should be done. Various techniques for saving of energy in a transformer which is already in service have been discussed. This chapter also highlighted about the material which should be used during the manufacturing stage to reduce the losses.

3.7 Software

After going through the basic details of the transformer, its working principle, various energy saving aspects, we would look into how the software supplied with this book can help us in our energy saving programme for the transformers.

3.7.1 Objective

- To create a data bank of information related with the transformers of a plant for which energy auditing is required.
- To calculate the load factor of transformer for known value of load losses and no load losses at which the efficiency will be maximum.
- To calculate efficiency of transformer for a known load factor and power factor.
- To evaluate transformer of different manufacturers for TOC (Total Ownership Cost).
- Printing of results of various calculations.

3.7.2 Procedure

- For preparing the data bank of transformers, click on "Electrical Equipment: Inventory" of the main menu, a submenu will appear. Click on "Transformer" of the submenu as shown by an arrow in Fig. 3.2. A screen will appear as shown in Fig. 3.3.

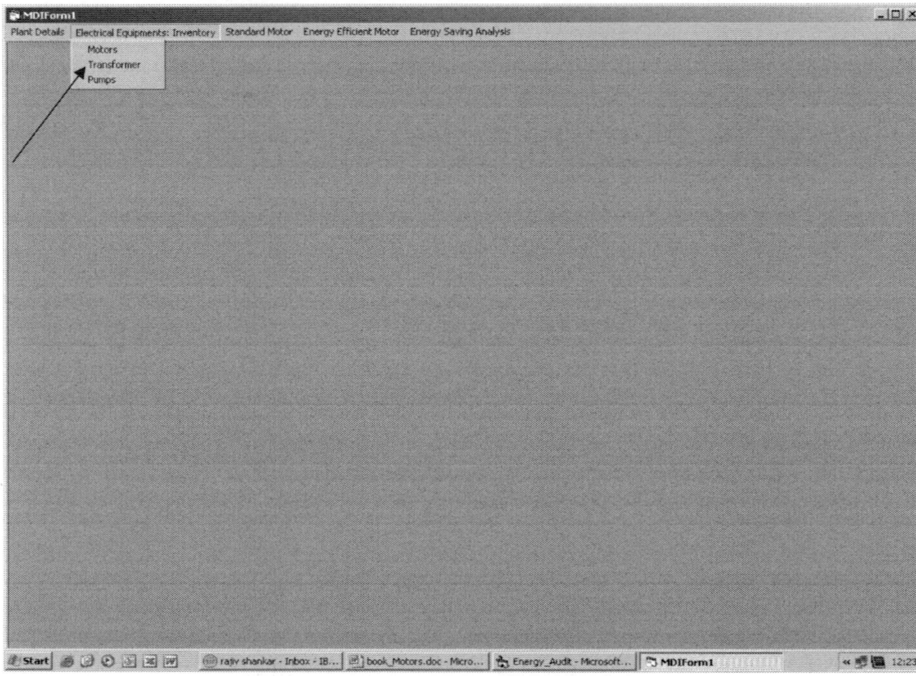

Fig. 3.2 Main screen

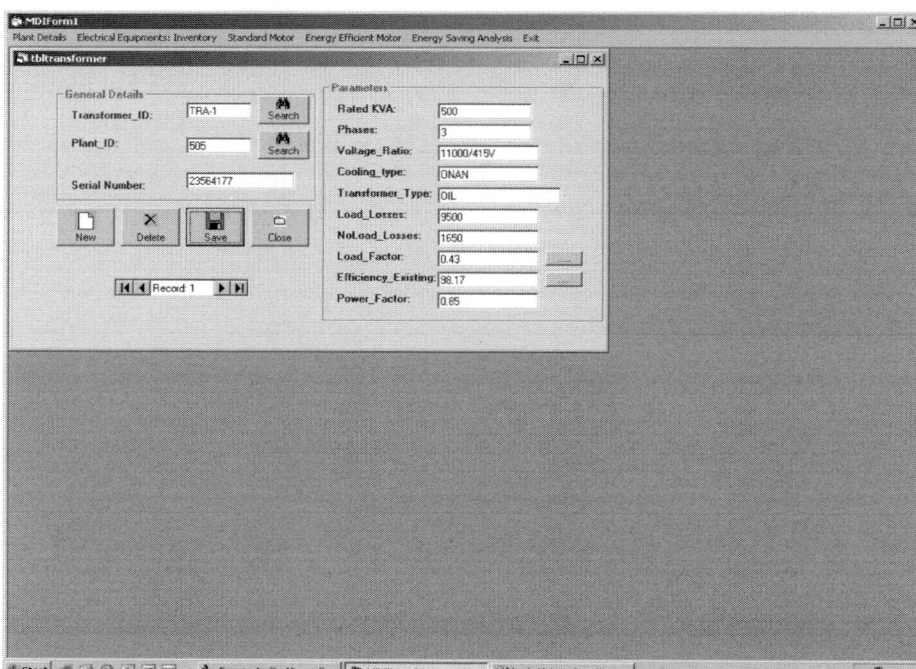

Fig. 3.3 Transformer data filling screen

- Fill up the details in blank boxes, which can be obtained either from the name plate or test certificate of transformer.

  Transformer-ID: This is a Transformer Identification Number. It can be a pure numerical or alpha numerical, like TR1 or TR12 etc. A search button is provided near the Transformer-ID box for editing any records available in the data bank, refer Fig. 3.4. Click on the button and a screen as shown in Fig. 3.5 opens up. Click on "Retrieve" button as shown by a dotted arrow in Fig. 3.5. Transformer-ID and KVA rating make all the transformer available in data bank will appear on the screen in respective columns of the grid.

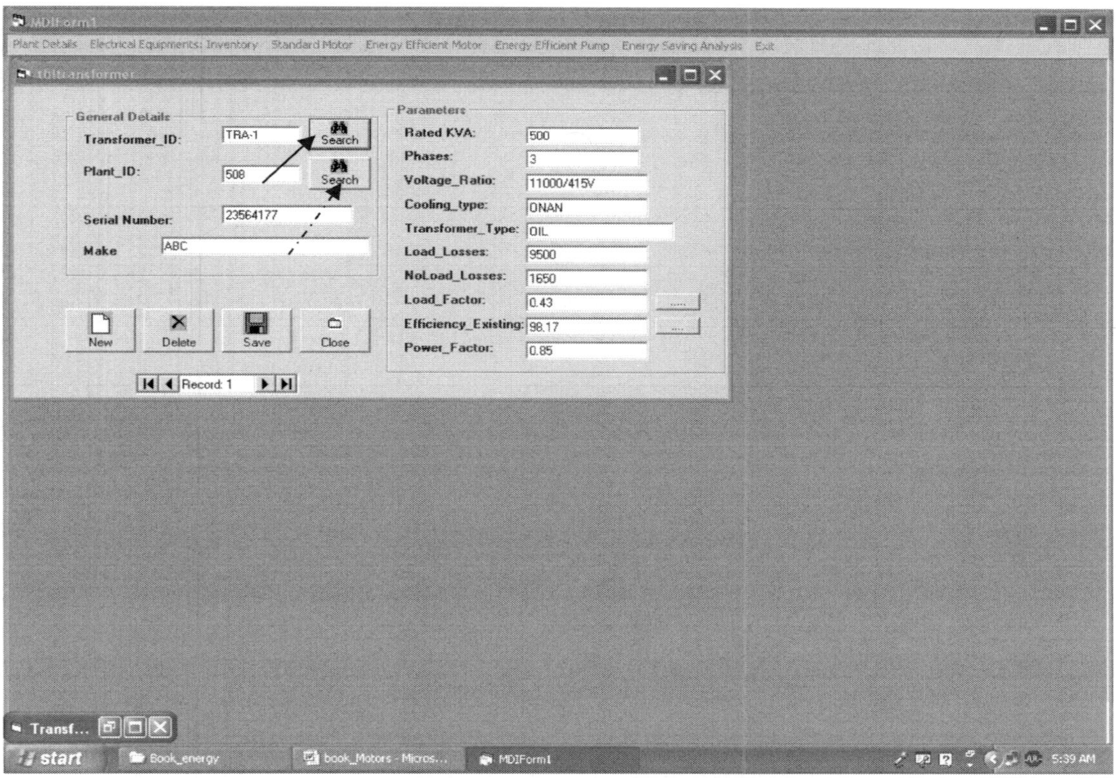

Fig. 3.4 Search button shown by an arrow

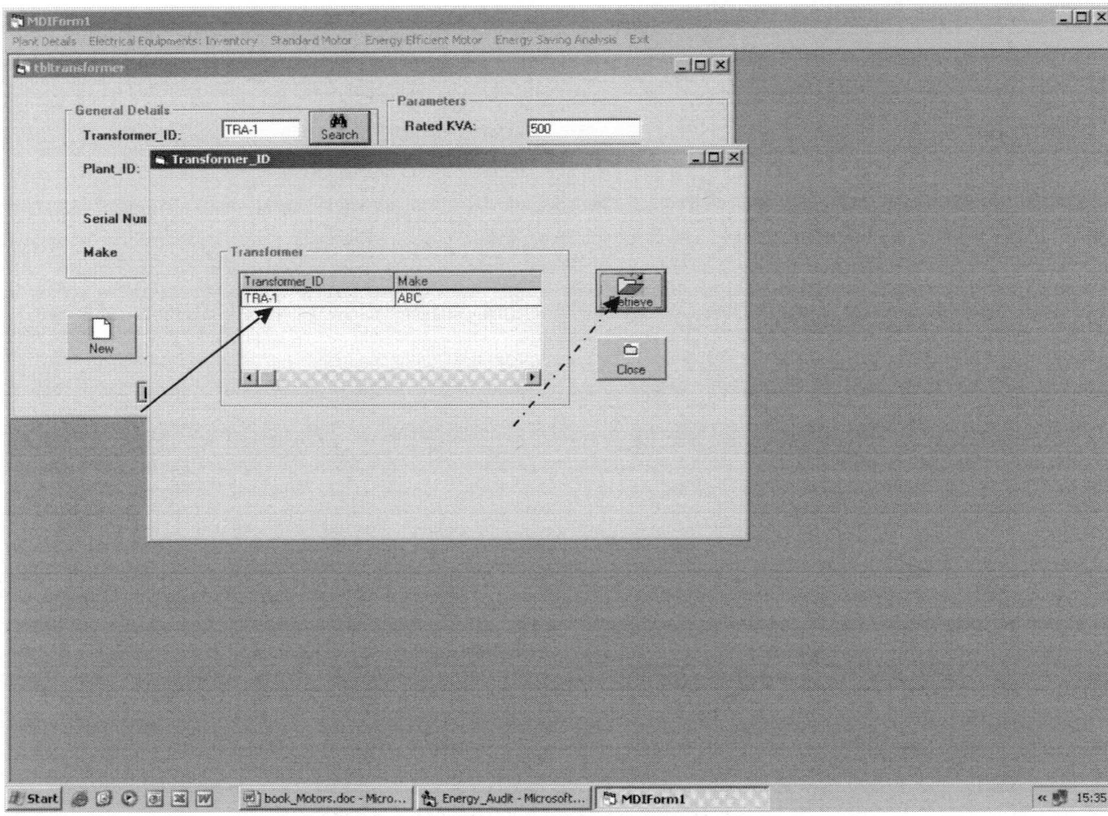

Fig. 3.5 Search for transformer records

For transferring any record to the main transformer screen, just click on the "Transformer-ID" as shown by an solid arrow in Fig. 3.5 and all the stored records will be automatically transferred to the main transformer screen. After transferring the records to the main screen editing can be done and then edited records can be saved by clicking on "Save" button.

Plant-ID: Plant-ID should be of that plant to which the transformer belongs. To find out Plant-ID, click on the "Search" button provided near the Plant-ID as shown in Fig. 3.4 by a dotted arrow. A Plant-ID screen will appear as shown

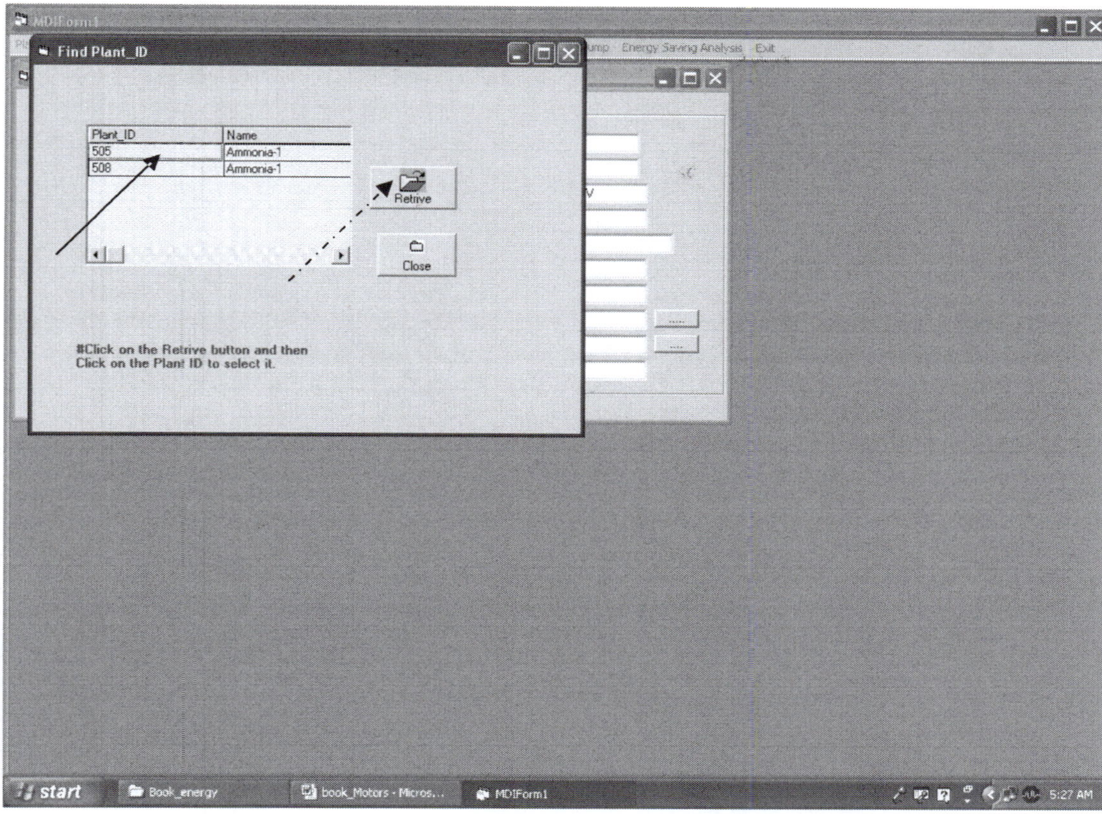

Fig. 3.6 Screen for searching Plant-ID

in Fig. 3.6. Click on "Retrieve" button as shown by a dotted arrow in Fig. 3.6, to retrieve all the stored Plant-ID along with the name of the plant. Copy the respective Plant-ID from Plant-ID screen to Plant-ID box on the main transformer screen.

Load Factor and Efficiency: To find out load factor and transformer efficiency of that load factor, click on the button provided near the load factor on main transformer screen as shown in Fig. 3.7 by a dotted arrow.

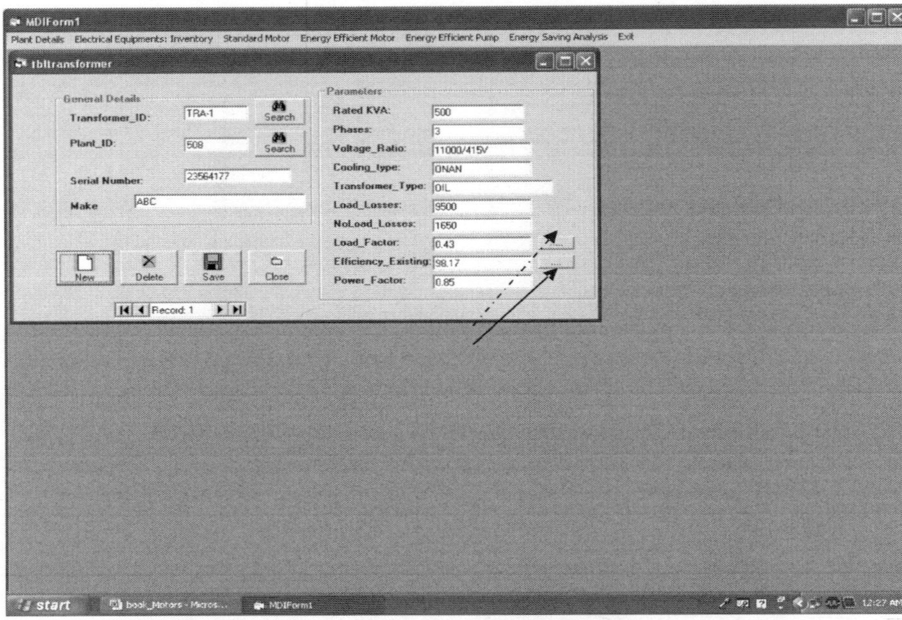

Fig. 3.7 Main screen showing load factor and efficiency button

On clicking, a new screen will appear as shown in Fig. 3.8. Fill up the required values and click on "Calculate" button to calculate load factor and efficiency.

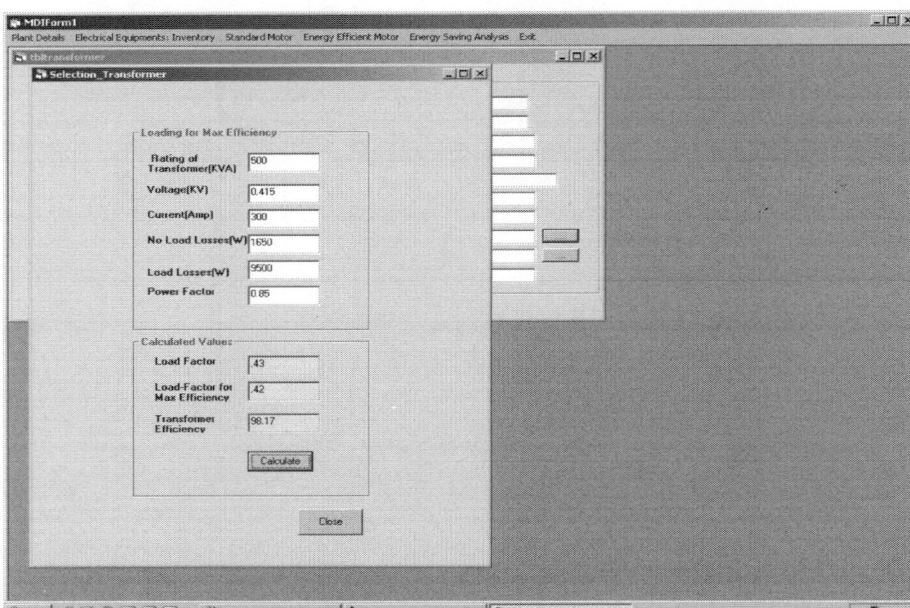

Fig. 3.8 Screen for calculation of load factor and efficiency

Let's take an example to further elaborate the above point.

Example 3.2

Following are the parameters of a 500KVA transformer.

KVA rating = 500 (as per name plate)
No load losses = 1650 watt (as per name plate)
Load losses = 9500 watt (as per name plate)
Current = 300A (measured)
Voltage = 0.415KV (measured)
Power factor = 0.85 (measured)

Load factor for maximum efficiency is required to be calculated.

Load factor for maximum efficiency
= Sq. root (no load losses/load losses)

Sq. root (1650/9500) = 0.416 or 0.42 approx

Existing load factor = $1.732 \times 0.415 \times 300/500$
= 0.43

Efficiency on the existing load factor:

Efficiency = Output/(Output + Losses)

$$= \frac{\text{Load factor} \times \text{KVA rating} \times \text{Power factor} \times 1000 \times 100}{(\text{Load factor} \times \text{KVA rating} \times \text{Power factor} \times 1000) + ((\text{load factor})^2 \times \text{Load losses} + \text{No load losses})}$$

$$= \frac{0.43 \times 500 \times 0.85 \times 1000 \times 100}{0.43 \times 500 \times 0.85 \times 1000 + ((0.432)^2 \times 9500 + 1650)}$$

= 18275000/186156.55
= 98.17%

Same result is obtained from the calculation by using the software as shown in Fig. 3.9.

- After building data bank of transformers, the next step is analysis of the transformers for energy conservation. If two or more than two transformers are provided in a sub-station, then the load is redistributed between the transformers in such a way that each transformer operates near to its maximum efficiency point. To simplify the case, we assume that the average load remain constant through out the period for which the transformers remain energized. As explained earlier that load factor of the transformer plays an important part in deciding the efficiency of transformer and at particular load factor, efficiency will be maximum. Therefore, load should be distributed in such a way, so that the efficiency is maximum. As explained earlier also, value of the load factor at which the efficiency will be maximum can be easily found out from the software. In some situation, it may be possible to completely switch off the lightly loaded transformer, so that the core losses will become zero, but in most of the cases due to criticality of the supply, it is not always possible to switch off the transformer.

- The next step is to analyse the saving in energy achieved by redistribution of load. Click on the "Energy Saving Analysis" of the main menu and then on the "Existing Transformer" of the submenu as shown in Fig. 3.9.

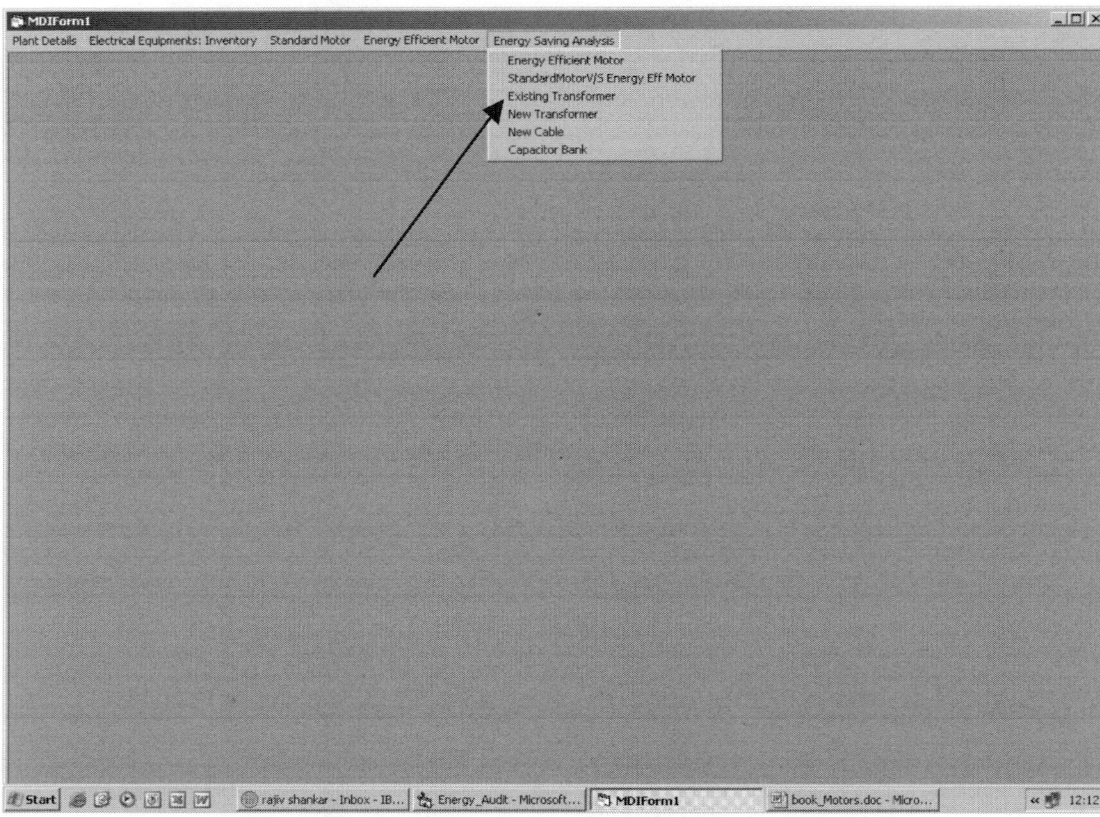

Fig. 3.9 Analysis for energy saving of existing transformer

Clicking action will open up a new screen as shown in Fig. 3.10.

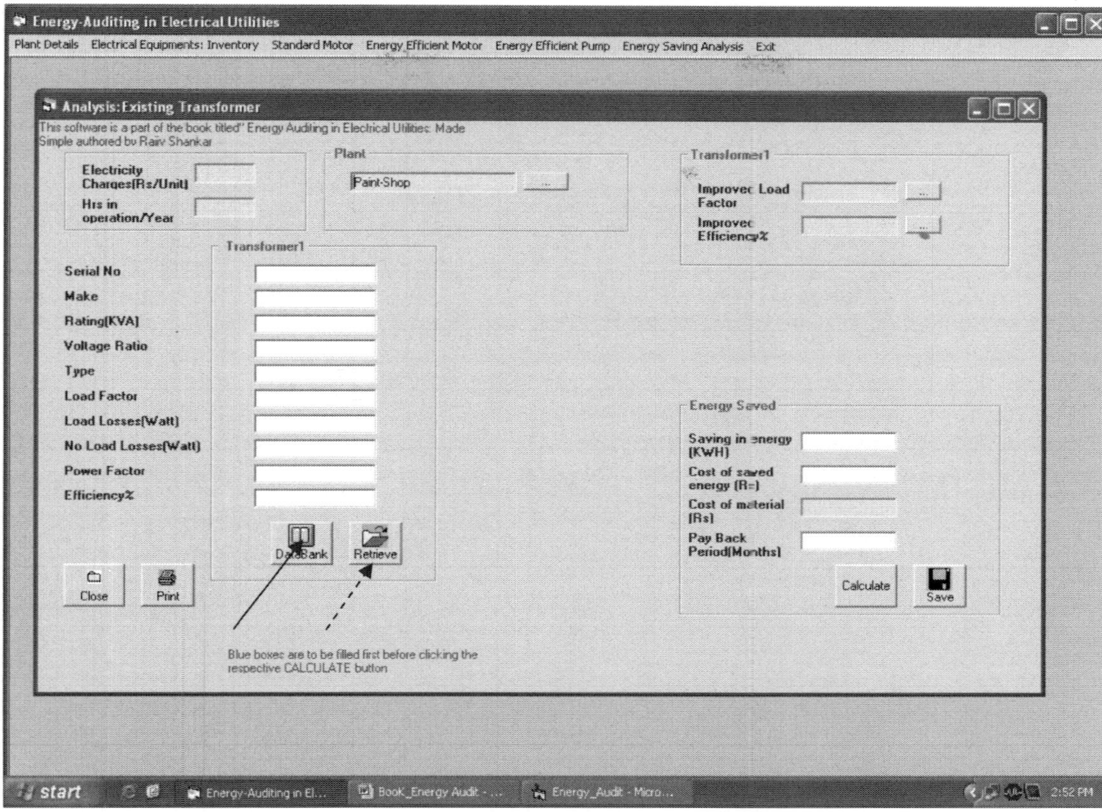

Fig. 3.10 Screen for analysis of existing transformer

Click on the "Data Bank" button as shown by a solid arrow to go to query screen as shown in Fig. 3.11. Retrieving of data from data bank which has already been created, is based on generation of query as explained in previous chapter. Query is generated by filling in the range of transformer KVA rating and existing efficiency of the transformer. Records which fall within the range of query parameters are selected from the data bank by clicking on "Search" button as shown in Fig. 3.11 by a dotted arrow. Click on "Transformer-ID" to transfer the records on the main screen for analysis and then click on "Retrieve" button to retrieve all the records as shown in Fig. 3.10 by a dotted arrow. The blank boxes will be filled by the retrieved values.

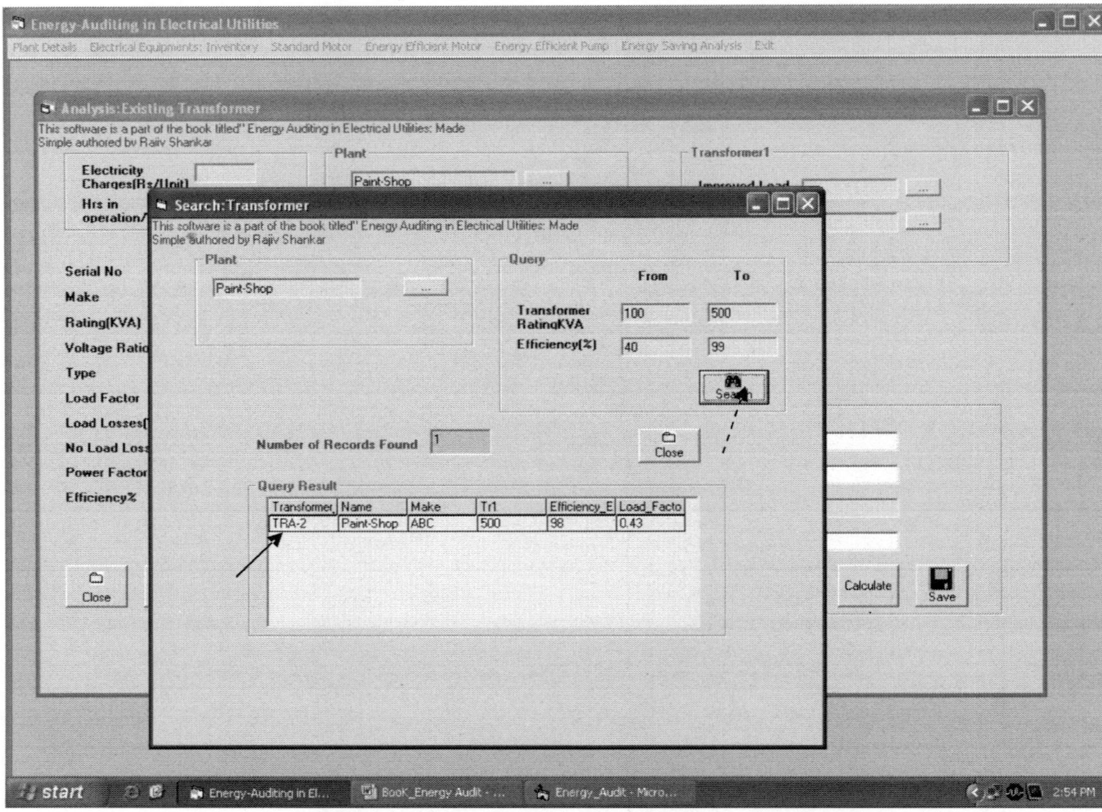

Fig. 3.11 Screen for generation of query

Obtain values of improved load factor and improved efficiency after clicking on the button provided near the load factor and efficiency field, refer Fig. 3.12. To calculate the amount of saved energy and the payback period, electricity charges, working hours of transformer and cost of material required to shift the load from one transformer to another are also needed to be filled in the blank boxes. Click on "Calculate" button to calculate the cost of saved energy and payback period. Click on "Save" button, as shown by a dotted arrow to save the results.

Click on "Print" button to print the results as shown in Fig. 3.13.

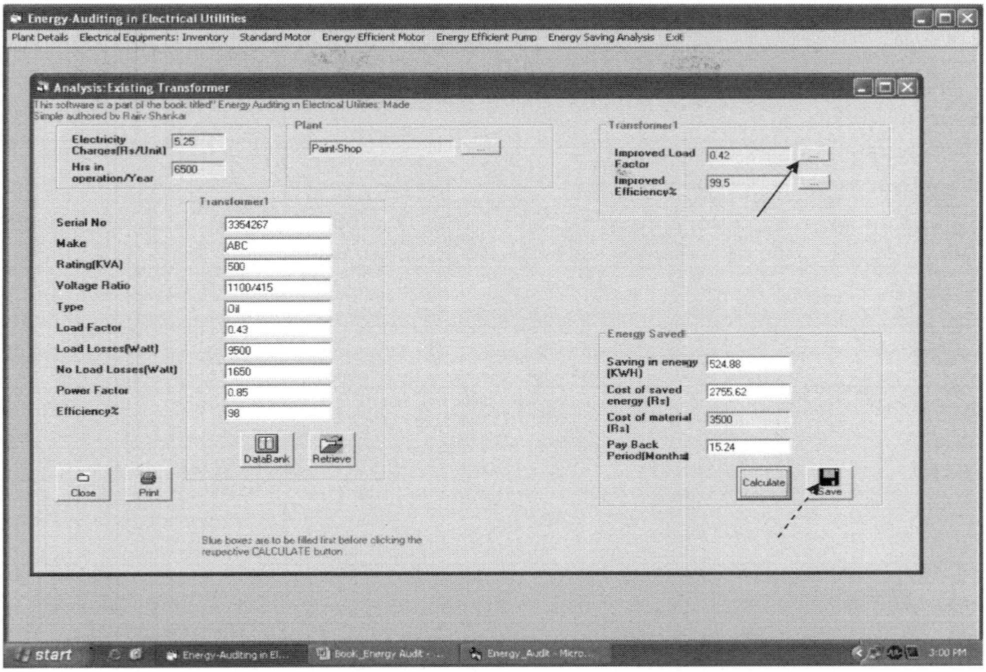

Fig. 3.12 Calculation of cost of saved energy and payback period

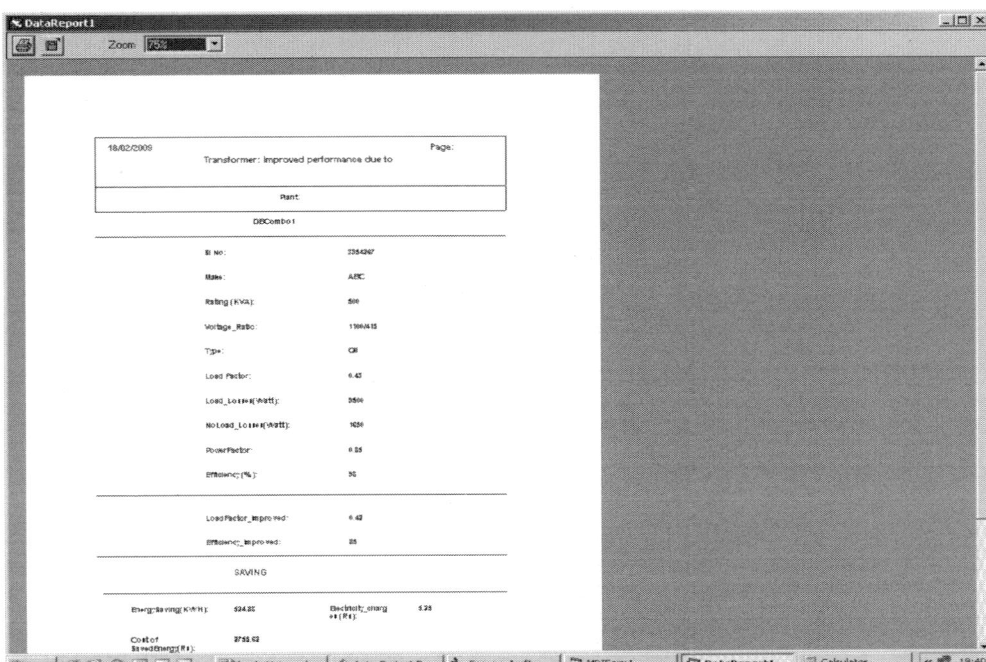

Fig. 3.13 Printout of the analysis result

Example 3.3

Transformer rating = 500KVA

Load losses = 9500 watt

Existing load factor = 0.43

New load factor after diversion of some of the load to other transformer = 0.42

Load losses at 0.43 load factor $= 0.43^2 \times 9500$

$= 1756.55$ watt

Load losses at 0.42 load factor $= 0.42^2 \times 9500$

$= 1676$ watt

Saving in losses $= 1756.55 - 1676$

$= 80.55$ watt

Loading period of transformer is 6500 hrs in a year (for simplicity, it is assumed that load remains constant).

Energy saved in a year $= 80.55 \times 6500 = 523.57$KWH

Cost of energy saved = Rs5.25 × 523.57 = 2748.77

Expenditure made in diversion of load = Rs3500

Simple payback period $= 3500 \times 12/2748.77$

= 15.27 or approximately 15 months

Similar results are obtained from the software also as shown in Fig. 3.13.

- As mentioned earlier also, for a new installation, the criteria for selection of transformer from different makes should be such that the transformer having less TOC should be selected.

  The software helps us in calculating the TOC.

- Click on "Energy Saving Analysis" on the main menu of main screen. On clicking "New Transformer" of the submenu as shown by an arrow in Fig. 3.14, a screen will appear as shown in Fig. 3.15.

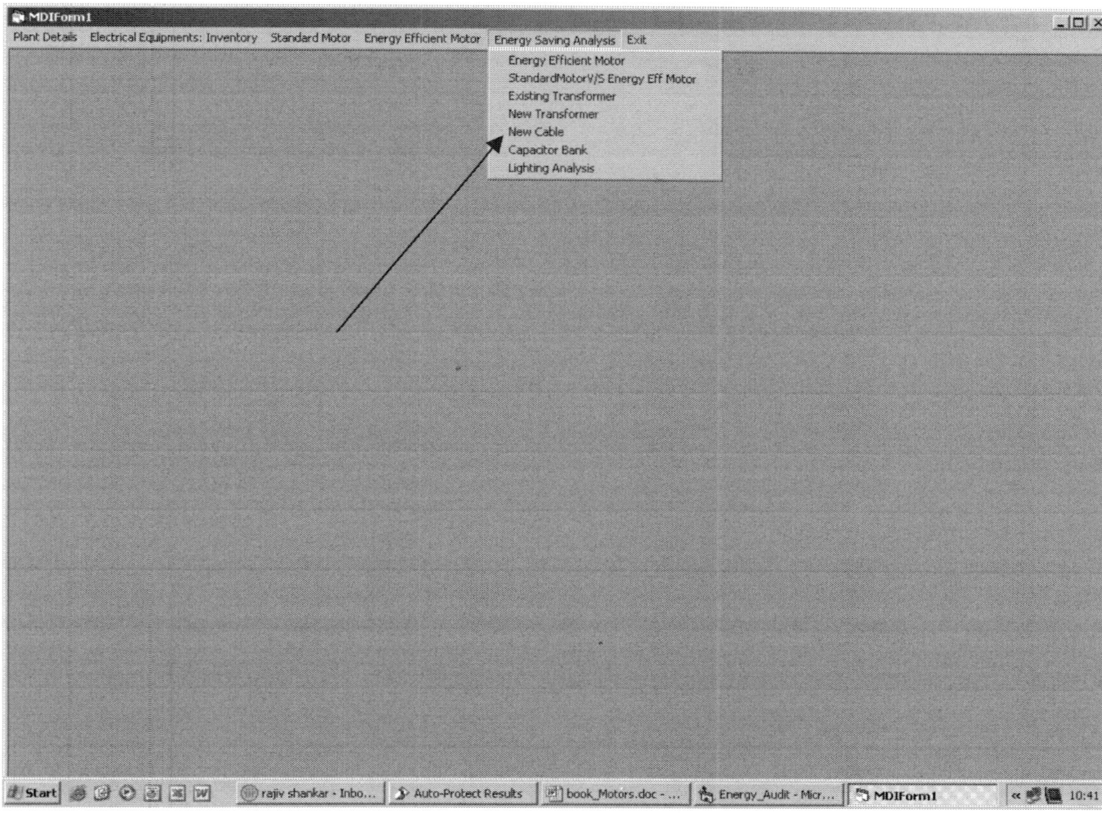

Fig. 3.14 Main screen

Fill in the common data like transformer loading hours, per unit electricity charges, loading factor and discount rate and life of the transformer as shown in Fig. 3.15. Click on the "Present Value" button to calculate present value per watt of the transformer losses.

The present value of load losses and no load losses of transformer can be calculated by clicking the "PV Cost of Load Losses" and "PV Cost of No Load Losses" button respectively. The total cost of the transformer that includes the capital cost of the transformer is calculated by clicking the "Total Cost of Transformer" button.

Same procedure is repeated for Transformer 2 also.

The results can be saved and printed by clicking the "Save" and "Print" button respectively as shown in Fig. 3.16.

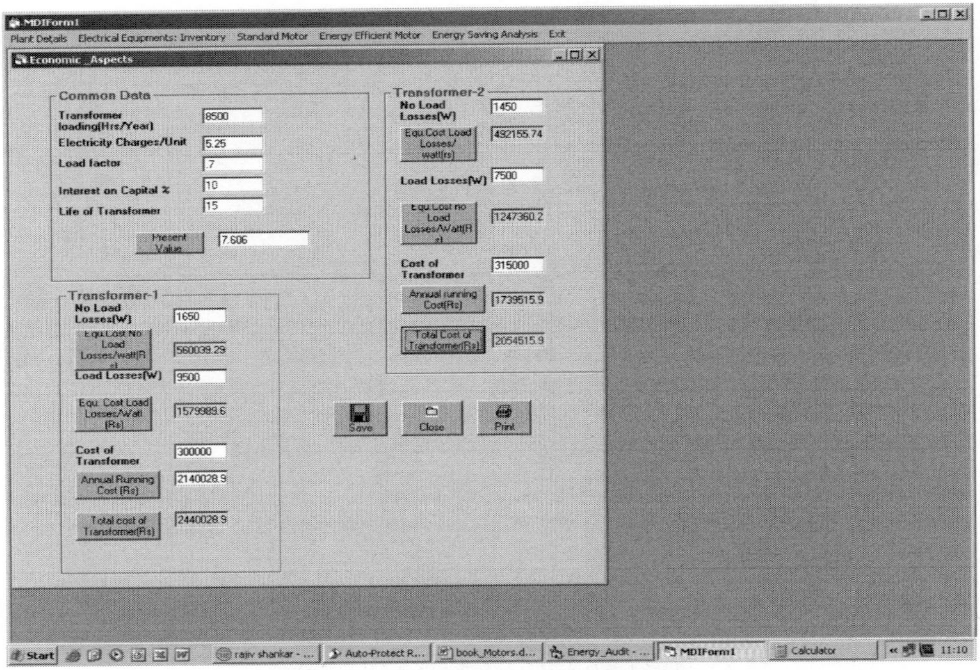

Fig. 3.15 Calculation of TOC of two transformer

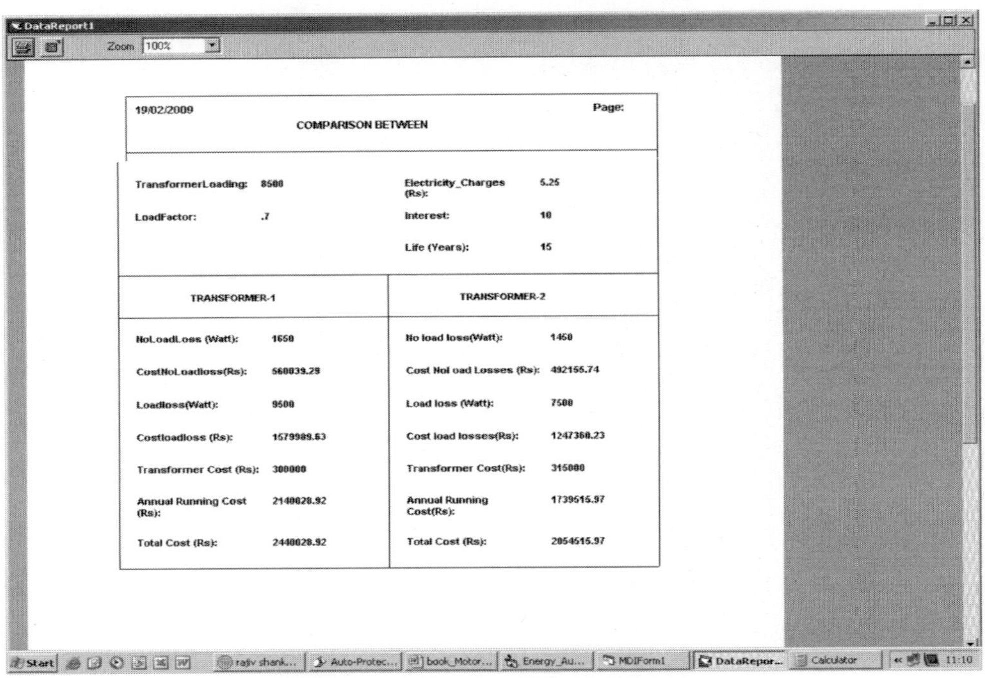

Fig. 3.16 Printout of the TOC of two transformers

CHAPTER

4

HT and LT Cables

4.0 Introduction

In an electrical system, cable is required to transmit power from one source to another. It can carry a small as well as large load. Different sizes of the cables will be required for carrying power at different voltages. Cables can be classified as low voltage and high voltage cables. Cable must be selected and installed in such a way that it should be capable to transmit power at the required voltage without damaging itself or the other connected equipments. This aspect will be discussed in more detail in this chapter. Cables must be installed in such a manner that it present no danger to the person who may come in contact with the cable. In high voltage cable, induction level should not be so high that it develops the hazardous conditions to the near by equipments. Cable should be selected and installed as per the environmental conditions of that area. For example, the temperature rating of the cable must suit the ambient condition.

Broadly, cable can be divided into two categories.

- Power cable
- Control cable

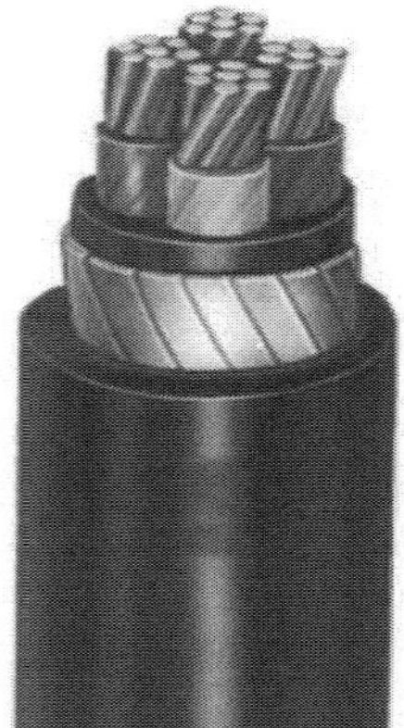

Fig. 4.0 Stranded flat armoured copper

Power cables are used to carry bulk power from one source to another and generally, the conductor is made of aluminium. Aluminium conductor will take 1.6 times more in cross section area to carry the same current. Therefore, aluminium conductor occupies more space in comparison of copper conductor.

The conductor material of control cable is always stranded or flexible copper. Control cables are always low voltage cable, since control supply is always a low voltage supply, while power cables up to 66KV are very common.

A wide variety of insulation is provided for the cables depending on the application and environmental conditions. Some of the insulating materials are rubber, PVC, XLPE and paper. For HT cable, XLPE insulation is used. But now-a-days, LT cable are also available with XLPE insulation. XLPE insulation is preferred over the PVC insulation since it has better properties than the PVC insulation. Refer Table 4.1, which compares the properties of PVC and XLPE insulation.

Fig. 4.1 XLPE cables

Table 4.1 Comparison Between XLPE and PVC Cable

S. No.	*Characteristics*	*PVC*	*XLPE*
1.	Conductor Temperature Rating	70°C	90°C
2.	Life		Longer than PVC
3.	Max. Temperature Limit under Short-circuit	160°C	250°C
4.	Over-loading	10%	60%
5.	Moisture Resistance		100 times more than PVC
6.	Insulation Resistance	high	1000 times of PVC
7.	Weight		Less than PVC

4.1 Selection of Cable

There are many aspects of cable which needs to be taken care of while selecting a cable for a particular application. Some of them are mentioned below:

- System voltage
- Type of system i.e. earthed or unearthed system
- Maximum load current which is required to be carried by the cable
- Short-circuit current and its duration
- Voltage drop during starting of motor in normal condition
- Future load requirement
- Installation conditions i.e. installation of cable underground, in duct or in open
- Ambient temperature
- Proximity with other cables
- Depth of cable in case of underground installation
- Installation of cable in hazardous area
- Statutory regulations
- Economic aspects

4.2 Construction

The most commonly used material for conductor is copper and aluminium. As mentioned earlier also that in power cable aluminium is generally used while in control cables copper is used. Copper has better conducting properties than aluminium. The aluminium cable needs larger surface area for termination. The termination of aluminium cable with the equipment having copper terminal causes bimetallic effect and formation of aluminium oxide on the metal surface. Using bimetallic washers/sheet in between solves this problem. Since copper is not available in sufficient quantity, aluminium cable is used in power circuits.

The conductor of power cable consists of aluminium of electrical grade purity and generally it is stranded and shaped conductor. Sometimes, it may be solid also. In high tension cable, the conductors are stranded, compacted and round. Shaped conductors are not preferred in high tension cable since the electrical stresses are higher at the edges of the HT cables. The control cable consists of annealed high conductivity copper. All conductor conforms to IS:8130. Normally, the aluminium conductor size is up to 10 sq.mm and are solid circular in cross section and above 10 sq.mm are stranded. Single and twin core cable is up to 50 sq.mm and are circular in shape while conductors of cross section area is 25 sq.mm and above are shaped in three and four core cables.

4.2.1 Insulation

A large variety of insulating material is used in cables, which depends on the application, installation and environmental conditions. For example, rubber insulation is provided for flexible moving cable. Other insulating material are PVC (Poly Vinyl Chloride), XLPE (Crossed Linked Poly Ethylene), Paper etc. For fixed installation, power cables with PVC and XLPE insulation are most commonly used. In HT cable, the XLPE insulation is provided in its natural colour since adding colour for core identification deteriorates the electrical properties. The insulation should be free from air voids and foreign material as presence of these material results in unequal distribution of electrical stress. The insulation and outer sheathing compound conforms to IS:5831.

Presently, XLPE insulated cable are preferred over the PVC in LT installation also since qualities of the XLPE insulating material are superior than PVC. The dielectric loss of XLPE insulation is very low and does not vary much over the entire operating range. XLPE insulated cable can carry much higher current for the same cross section area in comparison to the PVC insulated cable. Cable with XLPE insulation has also higher short-circuit with stand capacity than the PVC cable, since it can withstand higher temperature rise in comparison to PVC insulated cable. XLPE cable can be overloaded up to 60% in case of emergency due to failure of one cable where cables are installed in parallel and in this way total current carrying capacity of the system will not be hampered.

4.2.2 Inner Sheath

In LT cables a common covering is applied over the core of extruded PVC compound or by wrapping of plastic to provide bedding for the armour. In HT cables both conductor and insulation shall be provided with extruded semi-conductor compound shielding. The XLPE insulation is weak to tangential electrical stress and hence for better performance, electrical stress should be radial. The main purpose of providing the semi-conductor compound shield is to make the electrical stress radial. In addition to that HT cable insulation is provided with semi-conductor tape and non-magnetic metallic tape screen over the extruded insulation to reduce the electrical stress and to carry both leakage and earth fault current. Common inner sheath of extruded PVC is provided. Fillers of non-hygroscopic material are provided.

4.2.3 Armouring

The armouring consists of galvanized flat or round steel wire. For single core cable, non-magnetic hard drawn round or flat aluminium wire is provided.

4.2.4 Outer Sheath

The tough outer sheath of PVC compound is extruded over the armour/non-magnetic metallic part of insulation/inner sheath (unarmoured cable). In FRLS cables (Flame Retardant Low Smoke) special material is used in the outer sheath.

The following tests are performed on the FRLS cables.

- Smoke density measurement
- Oxygen index test
- HCL liberation test

4.3 Specification

While ordering cable, following parameters are required to be specified.

- System voltage
- Core size and number of core
- Cable is required for grounded or not grounded system
- Conductor material
- Type of armour i.e. flat or round
- Type of insulation
- Relevant IS number
- Length of cable in one drum
- Ambient conditions
- Any other special requirement

4.4 Tests

Following routine tests are carried out on a cable.

- Conductor resistance test
- High voltage test
- Partial discharge test (in case of HT cables)

For high voltage test, the test voltages are shown in Table 4.2 is applied between conductor and screen/armour.

Table 4.2

Voltage Rating	*Test Voltage*
1.1KV (E)	3KV
3.8/6.6KV (E)	12KV
6.35/11KV(E)	17KV
11/11(KV)(UE)	28KV
12.7/22KV(E)	32KV
19/33KV(E)	48KV

The above voltage is applied for five minutes.

4.5 Installation

Installation of cable shall be done as per IS:1255. If installed in air, cable trays of adequate size shall be provided. Cables shall be properly clamped on cable trays. If the cables are installed in group and the distance between the two cables is 75mm (in a horizontal plane) no de-rating is required up to 70 sq.mm cable and above 70 sq.mm the distance should be 150mm. If it is less, then de-rating is required. The bending radius should be 15 times the diameter of the cable. In case of 1.1KV multi core cable, it can be 12 times of the diameter of the cable. Further, as mentioned above cables are selected as per the environmental conditions. The current ratings are based on the normal/standard conditions of installation as mentioned below.

	PVC Cable	*XLPE Cable*
Maximum Conductor Temperature	70°C	90°C
Ambient Air Temperature	40°C	40°C
Ground Temperature	30°C	40°C
Depth of Laying	75cm	90cm
Thermal Resistivity of Soil	150°C cm/watt	150°C cm/watt

If the cables are installed in such a way that it is exposed to the sun then the rating of the cables will get reduced in comparison to the cables, which are shielded from the sun.

Short-circuit ratings of the cables are specified by the manufacturer for a temperature rise of 250°C (XLPE) and 160°C (PVC) for one second. If the cables are subjected to more than that period, the value as specified by the manufacturer should be divided by $\sqrt{t}$, where, t is the time for any other short-circuit duration.

Proper care should be taken while terminating the cable.

- Correct size of lug and also of correct material shall be used. Crimping of lugs must be done by the proper crimping tool.
- Sufficient clearance should be maintained in the terminal box where cable is to be terminated.
- Nut-bolts and spring washers of adequate size should be used for properly tightening the lugs to the terminal strip. Looseness of contact will result in heating and burning of cable insulation, terminal strip and sometime it results in short-circuiting.
- Segregation of cable should be maintained i.e. power cable, control cable and communication cable should be separated from each other.
- If cables are bunched together then it should be de-rated by applying appropriate de-rating factor.
- Cable tags and core identification numbers are provided which will be helpful in maintenance of the electrical system.
- If cables are laid underground then sand for cushioning effect and bricks for mechanical protection should be provided. Hume pipes of proper size shall be provided where cables are supposed to cross the road, drain etc.
- Cable route marker at appropriate distances shall be provided for underground cables.

4.6 Economics in Selection of Cable

As mentioned above that the conductor of cable is either copper or aluminium and both of these have got some resistance. Whenever a cable carries current, due to the resistance of the cable, resistive losses take place. These losses are dissipated in the cable in the form of heat.

The resistive losses depend on the square of the current and the value of the resistance of the conductor.

We don't have much control on the current since it depends on the load. However, if we select a cable of higher size than the calculated size, the resistance will be lower since resistance is inversely proportional to the cross section are a of the conductor and the cable of higher size has more cross section area. At the same time the cost of higher size cable will also be more. Therefore, an economic model should be worked out if higher size cable is selected. The calculation for life cycle losses shall be done as explained in earlier chapters. The ownership cost of the cable should be worked out by adding capital cost or the procurement cost and the present value of cost of losses over the life of the cable.

IS standards:
PVC cables: IS:1554 (Part I & II) 1.1KV, 3.3KV, 6.6KV, 11KV
XLPE cables: IS:7098 (Part I) 1.1KV(E)
IS:7098 (Part II) 3.3KV, 11KV, 22KV, 33KV
EHV XLPE cables: IEC 60840, IEC 62067
66KV to 400KV

4.7 Conclusion

Sometimes, it is cheaper to procure higher size of cable, since the resistance of higher size of cable will be less, the losses which depend on the resistance of the cable will also be less. Therefore, proper cost of evaluation is required while selection of cable is made for a particular application.

4.8 Software

The major loss which takes place in the cable is I^2R loss. This loss varies as square of the current and also depends on value of the resistance. The magnitude of the current passing through the cable depends on the load and varies according to the load. Resistance of the cable depends on the conductor size and the material. If we choose a cable of one size higher for a particular load the losses will be lower, however the cost of the cable will be more but the cost of losses in the cable will be less. Therefore, it is required to calculate the present value of the cost of the losses which take place in a cable while carrying current and a comparison should be done between the two cable i.e. cable which we consider is appropriate for a particular load as well as cable which is one size higher. Let's see how the software helps us in this regard?

4.8.1 Objective

- Calculation of present value of the cable losses.
- Calculation of total ownership cost of cable.
- To make a comparison between two sizes of cable.
- Printing of the calculated result.

4.8.2 Procedure

- Click on "Energy Saving Analysis" of main menu, a submenu will appear, click on "New Cable" of submenu. It is shown by an arrow in Fig. 4.2. A new screen will appear as shown in Fig. 4.3. The details of the cable like length of cable, loading hours per year, load on cable, per unit charges of electricity, no. of phases, interest/discount rate and life of the cable are to be filled in the blank boxes.

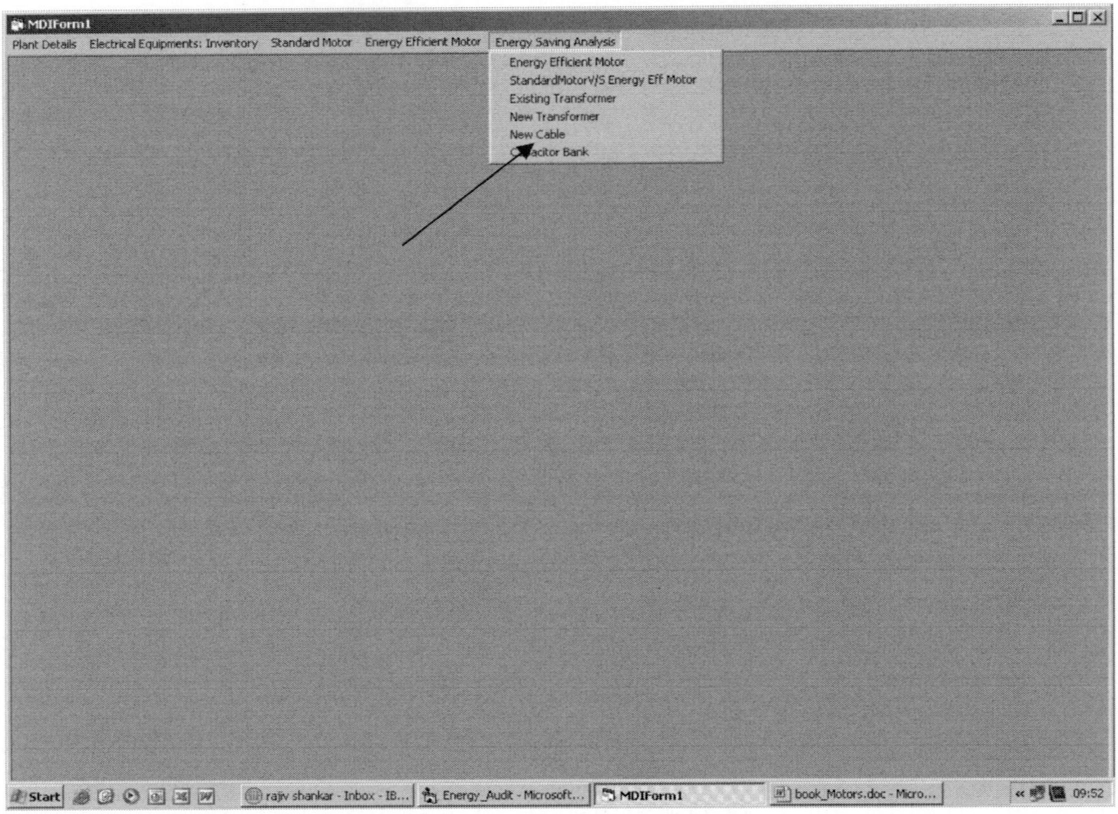

Fig. 4.2 Main screen

- Click on "Present Value" button to calculate the present value of the cable losses per watt.
- Fill in the cost and size of the cable in the respective boxes.
- To calculate the resistance of cable at the required temperature, click on "Calculate" button as shown by an arrow in Fig. 4.3. A screen will appear as shown in Fig. 4.4.

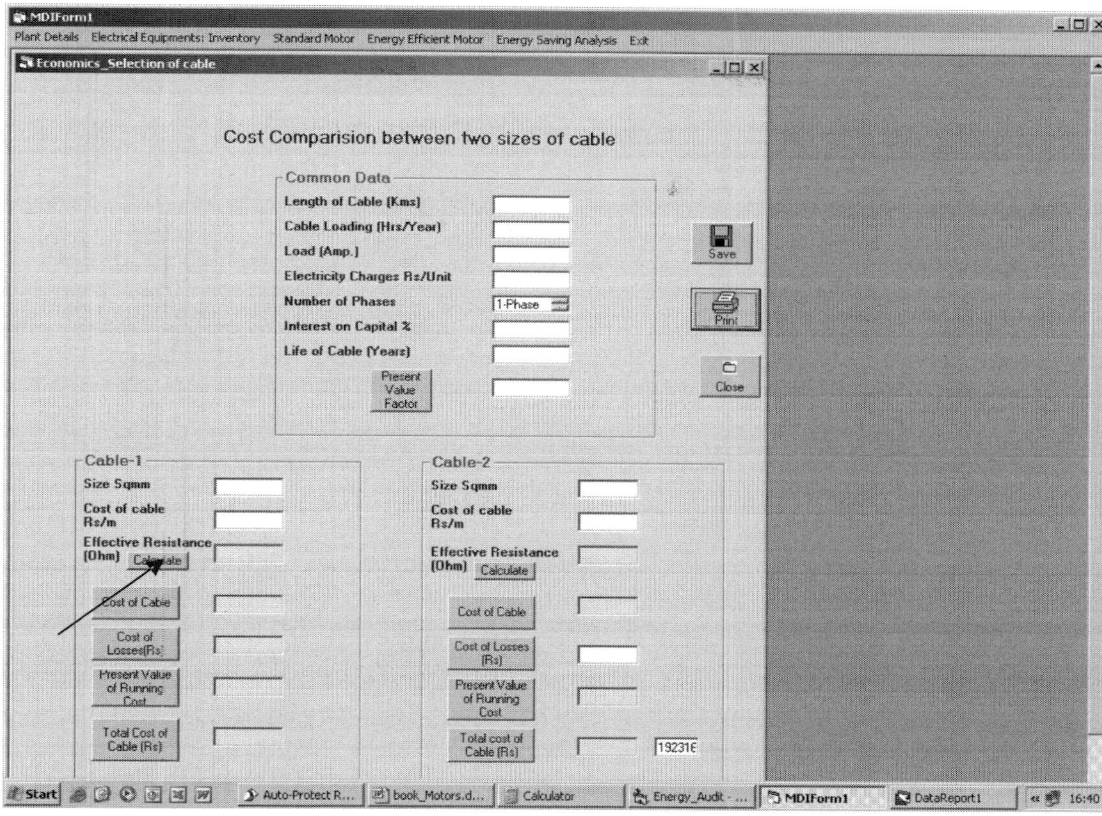

Fig. 4.3 Screen for analysis of cable

Fill in the respective boxes and click on the "Calculate" button to calculate the effective resistance at the conductor temperature.

After calculating the effective resistance, the other parameters like cost of cable, present value of cost of losses and total ownership cost is calculated. Similarly, total ownership cost of higher sizes of cables are also calculated and compared. The cable of lowest cost is selected.

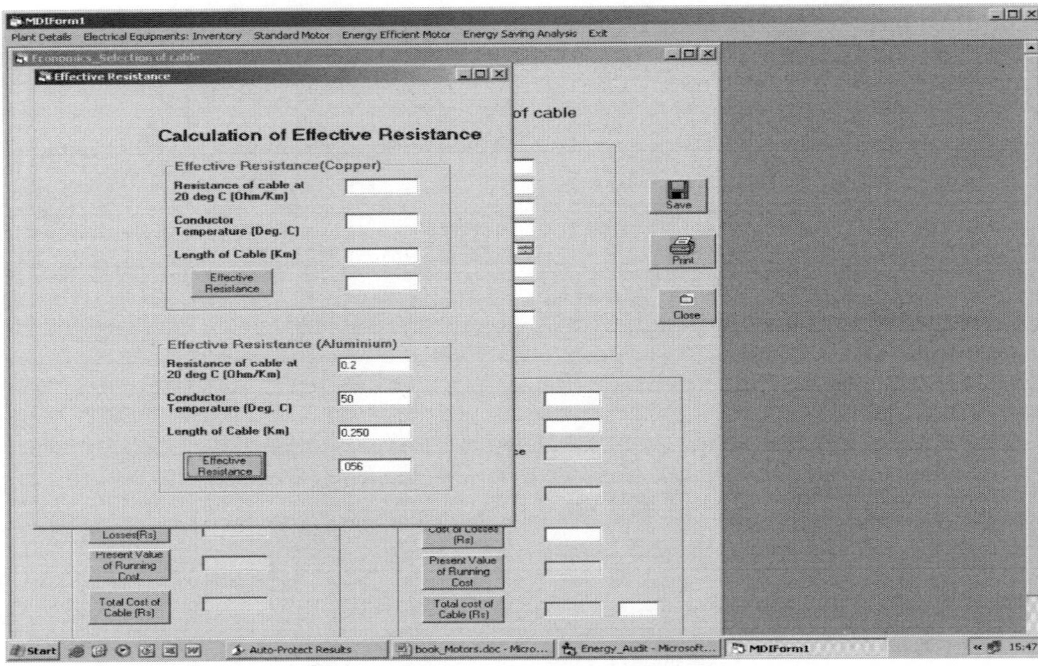

Fig. 4.4 Calculation of effective resistance

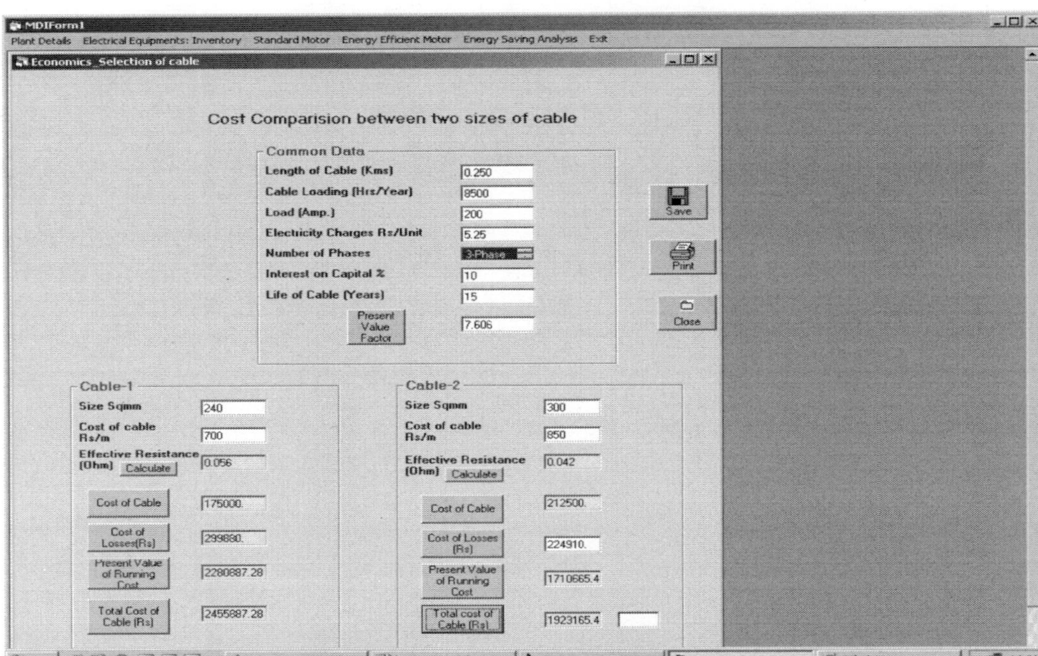

Fig. 4.5 Calculation of total ownership cost

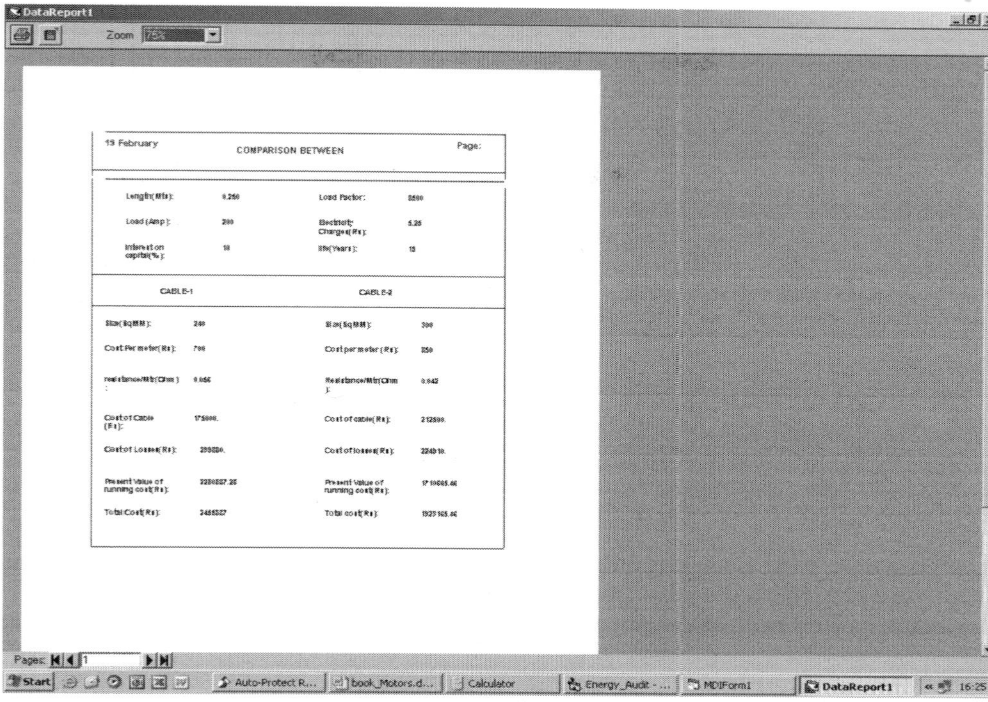

Fig. 4.6 Printout of analysis

- Click on "Save" button to save the calculation results. Results can also be printed by clicking on "Print" button as shown in Fig. 4.6.

  To clarify the above mentioned aspect, let's take an example.

Example 4.1

Let's take the following common parameters.

Length of cable = 250 metres

Loading hours of cable in one year = 8500 hours

Load on cable = 200 amp. (we have assumed a constant load throughout the loading period)

Core of cable = 3phase

Electricity charges = Rs5.25

Interest on capital = 10%

Life of cable = 15 years

Calculation of present value of a loss per watt

$$= ((1 + 0.1)^{15} - 1)/(0.1)(1 + 0.1)^{15}$$

$$= 7.606$$

Cable 1

3 core 240 sq.mm, aluminium conductor cable

Resistance of cable at 20°C = 0.2 ohm per Km

$R_1 = R_0 [T_1+225]/[T_0+225]$

R_1 = Resistance to be calculated at temperature T_1

R_0 = Resistance at temperature T_0

Resistance at 50°C

= 0.2 (50 + 225)/(20 + 225)

= 0.2245 ohm

Resistance of 250 metres of cable

= 0.2245 × 250/1000

= 0.056 ohm

Cost of cable per metre =Rs700

Cost of 250 metre cable

= 250 × 700

= Rs175000

Cost of losses

= $3 \times 200^2 \times 0.056 \times 8500 \times 5.25/1000$

= Rs299880

Present value of these losses

= 7.606 × 299880

= Rs2280887

Total ownership cost of cable i.e. purchase price plus present value of cost of losses

= 175000 + 2280887

= **Rs2455887**

Cable 2

3 core 300 sq.mm, aluminium conductor cable

Resistance of cable at 20°C = 0.15 ohm per Km

Resistance at 50°C

= 0.15 (50 + 225)/(20 + 225)

= 0.1684 ohm

Resistance of 250 metres of cable

= 0.1684 × 250/1000

= 0.042 ohm

Cost of cable per metre = Rs850

Cost of 250 metre cable

$= 850 \times 250$

= Rs212500

Cost of losses

$= 3 \times 2002 \times 0.042 \times 8500 \times 5.25/1000$

= Rs224910

Present value of these losses

$= 7.606 \times 224910$

= Rs1710665

Total ownership cost of cable i.e. purchase price plus present value of cost of losses

$= 212500 + 1710665$

= **Rs1923165**

From the above example it is clear that though the initial purchase price of higher size cable is more but if the present value of losses are also taken into account, the cable of higher size becomes cheaper than the cable of lower size.

CHAPTER

5

Lighting

5.0 Introduction

Lighting is an essential part of any commercial and industrial establishment and it plays an important role in increasing the productivity and reduction in accidents. IS:3436 specifies the level of lighting for different type of establishments. Nowadays, sophisticated software are available for designing outdoor as well as indoor lighting. Energy conservation measures in lighting system should be taken care of right at the time of designing of the building or industrial shed. The foremost requirement is that the design should be such as that during day time sufficient natural light should be available inside the building.

In this chapter, we will basically look into the aspects of energy conservation in lighting system.

Let's have a look at certain terms and definition most frequently used in a lighting system.

5.1 Terms and Definition

5.1.1 Lumen

Lumen or luminous flux is the measure of light output. A sphere of one metre radius, having a light source of one candela intensity in its centre, luminous flux falling on each square metre of the surface, is called One Lumen.

5.1.2 Lux

It's measure of illuminance of a surface. One lux is equal to one lumen per square metre. Sometimes, the terms lux and lumen appears to be confusing. Area is taken into account in case of lux, for example, a light source of 500 lumen falls on an area of 10 sq.m, then the illuminance of that area will be (500/10) 50 lux.

5.1.3 Load Efficacy

The unit of load efficacy is lux per watt per square metre. It is the value of average illuminance on a horizontal working plane per watt and per square metre.

5.1.4 Lamp Circuit Efficacy

It is the measurement of amount of lumens emitted by the lamp, for per unit of power consumption.

5.1.5 Colour Rendering Index (CRI)

It is the measure of effect of light on the colour of the object in a more natural form. Quantification of the ability of a light source to render colour of surface accurately is colour rendering index. This index is based on accuracy of a light source to reproduce a set of test colours relative to a test light source. A perfect match will have CRI of 100.

5.2 Characteristic of Different Types of Lamps

Table 5.1 Different Types of Lamps Commonly used in Domestic, Industrial and Various Types of Luminaries.

Lamps	*Lamp Wattage Including Ballast Wattage*	*Efficacy, Including Ballast, Lumen Per Watt*	*Colour Rendering Index*	*Lamp Life Burning Hours*
Incandescent Lamp	15, 25, 40, 60, 75, 100, 150, 200, 300, 500	8 to 17	100	1000
Fluorescent Light	18, 20, 36, 40	30 to 80	65 to 77	5000 to 15000
Halogen	200, 300, 500, 750, 1000, 1500, 2000	16 to 23	>90	2000 to 4000
Compact Fluorescent Lamp	8, 12, 15, 28, 32, 45	26 to 60	80	8000 to 10000
High Pressure Mercury Vapour Lamp	80, 125, 250, 400, 1000, 2000	50 to 60	45 to 60	5000 to 8000
Metal Halide Lamp	70, 150, 250, 400, 1000, 2000	50 to 70	70	8000 to 12000
High Pressure Sodium Lamp	70, 150, 250, 400, 1000	80	25 to 60	12000 to 20000
Low Pressure Sodium Lamp	35, 55, 135	90 to 130	40 to 60	12000 to 24000

5.3 Aspects of Lighting System Designing

For designing a lighting system, first of all lux level requirement is established. It will vary as per the location and activity. For example, lux level for street lighting and office lighting will be different. IS:3646 describes the illuminance requirements for various activities. IES (Illuminating Engineers Society) has also published recommended illumination level for various tasks.

Illumination levels for some of the activities as recommended in IS:3646 are mentioned in Table 5.2.

Table 5.2

Types of Interior of Activity	*Range of Service Illuminance in Lux*	*Remarks*
General Office	300-500-700	
Computer Work Station	300-500-750	
Conference Room, Executive Offices	300-500-750	
Bank Counter, Office Area	300-500-750	
Public Area, Banks	200-300-500	
Auditorium	50-100-150	Dimming facilities will be necessary. Special lighting of the aisles is desirable.
Hotel		
Entrance Hall	50-100-150	
Reception, Cashier's and Porter's Desk	200-300-500	Localized lighting may be appropriate.
Bars, Coffee Base dining-room, Lounges, 50-200		The lighting should be designed to create appropriate atmosphere.
Bedroom	30-50-100	Supplement local light at the bed head, writing table should be provided.
Bathroom	50-100-150	Supplementary local lighting near the mirror is desirable.

While designing an industrial shed, domestic or commercial establishment, the aspect of energy conservation in lighting system shall be given due consideration. The best way to conserve energy in lighting system is to make use of natural light as far as possible. Large size windows in office rooms, transparent doom in atrium of hotel lobby, transparent shed of a factory are some of the examples of use of natural light. More light should be provided at the work area in comparison of the general area. Fulfillment of above mentioned requirements is possible at the designing stage only.

Keeping the above mentioned aspects in mind, an estimate of number of luminaries is made. The first step in estimating the number of luminaries is to calculate the room index.

- Calculation of room index.

Room index can be calculated from the following formula:

$$\text{Room index} = \frac{L \times W}{(L + W) \times L_m}$$

L = Room length (metres)
W = Room width (metres)
L_m = Height of the luminaries above the working plane (metres)

The room index is normally 0.75 to 5.

Example 5.1

Assume that the length and width of room is 10 metres and 8 metres respectively. The height of the luminaries above the working plane is 2 metres then the room index is.

$$\text{Room Index} = 10 \times 8/(10 + 8) \times 2$$
$$= 80/36 = 2.22$$

The room index is required to find out the utilization factor. Utilization factor is the percentage of light (lamp lumen) reaching to the working surface after getting it reflected from the wall, ceiling and floor surfaces. Utilization factor tables are supplied by the manufacturer. The value of utilization factor is between 0 and 1.

Typical reflectance values are as per in Table 5.3.

Table 5.3 Typical Reflectance Values

	Ceiling	*Walls*	*Floor*
Air-conditioned Office	0.7	0.5	0.2
Industrial Building	0.5	0.4	0.2

- Using the room index and reflectance value utilization factor can be calculated.

Table 5.4 Utilization Table for a Typical Fitting

Reflectance			*Room Index*								
Ceiling	Wall	Floor	0.75	1.00	1.5	2.0	2.5	3.0	3.5	4.0	4.5
0.5	0.4	0.2	0.3	0.36	0.45	0.53	0.57	0.59	0.62	0.65	0.69

In the above table, where the horizontal row of reflectance intersects the vertical column of room index, that value will be the utilization factor.

For example, the room index as calculated above is 2.22 and the horizontal row of reflectance for a industrial building intersects the room index column 2.5 at 0.57, therefore, utilization factor will be 0.57.

- The next step is to calculate the number of fittings required for a room. Number of fittings required can be calculated from the following formula:

  N = F × A/L × Uf × Ma

  F = Lux level required

  A = Area of working surface in square metre

  L = Total lux of lamps in one lighting fixture

  Uf = Utilization factor

  Ma = Maintenance factor or light loss factor which takes into account loss of light due to dirt accumulation and also due to aging.

  Typical values of light loss factors are

  Air-conditioned office = 0.8

  Clean industrial environment = 0.7

  Dirty industrial environment = 0.6

  Area = 10 × 8 = 80 sq.m

  Lux level required = 200

  Lumen of one light fixture = 4000 (typical value)

  Utilization factor = 0.57 (as calculated above)

  Light loss factor = 0.7

  N = 200 × 80/(4000 × 0.57 × 0.7)

  = 16000/1695

  = 10 numbers

After calculation of number of light fixture, space, height ratio of the light fixture should also be checked. It should be as per the recommendation of the manufacturer of luminaries.

5.4 Installed Load Efficacy Ratio

Installed load efficacy ratio is a scale to ascertain achievement of energy efficiency from newly installed energy efficient lighting fixture or lamps in place of old fixture or lamps.

Installed load efficacy ratio shall be calculated as per the below mentioned steps.

- Find out the floor area.

- Measure the total power required to illuminate that floor area in watt.
- Calculate watt per square metre.
- Measure lux level by lux metre.
- Calculate lux per watt per square metre.
- Find out the means to increase the lux/watt/m^2 by replacing the energy efficient light fixture, use of natural lighting etc.
- Calculate the ratio of old lux/watt/m^2 and new lux/watt/m^2. This ratio is called Installed Load Efficacy Ratio.
- Calculate the annual energy wastage by using the following formula:
  (1-ILER) × watts × numbers of operating hours
  ILER = Installed load efficacy ratio

5.5 Various Means for Energy Saving

5.5.1 Use of Natural Daylight

Use of daylight reduces the energy requirement for lighting system to a considerable extent in commercial complex, industrial shed or public utility building. Various materials like transparent or translucent FRP can be used in roof of the building, industrial shed and also for manufacturing atrium doom etc.

- Special type of glass or glass strip running continuously across the width of the roof at regular interval can be used for natural daylight.
- Large size of glass windows can be provided for natural light in office rooms.

There may be various other options available to increase the daylight in the office or industrial premises. However, implementation of this concept is possible at the design stage of the building. Therefore, the architect of the building plays a major role in designing a building which has more energy conservation aspects.

5.5.2 Reduction in Light Fixture

A timer to switch off every alternate street lighting pole can be installed to reduce the number of lighting fixtures during night hours for a certain period of time when the tariff is very low. It will result in energy saving. Further, number of light fixture can also be reduced by providing efficient reflectors to enhance the luminance. Reflection of light can also be improved by painting the wall, ceiling in such a way that more light is reflected. Similarly, the floor tiles should be such as reflection of light will be more.

Another method of reduction in light fixture is to provide task lighting i.e. more lights are provided where task is being performed while the lighting level in general area is kept low.

5.5.3 High Efficiency Lamps and Luminaries

Colour rendering index of lamps should be considered for illumination of the area which is colour sensitive like assembly line, garment shop, painting shop. If we look into the Table 5.1, GLS lamps and halogen lamps has colour render index 100 and more than 90 respectively and these lamps are more suitable for colour sensitive areas, however the efficacy of these lamps are very low. Therefore, number of luminaries required will be more and power consumption will also be high. Metal halide lamp has less colour rendering index in comparison to the GLS and halogen but its efficacy is very high. The sodium vapour lamp has high efficacy in comparison with metal halide lamp but colour rendering index is low. Therefore, the metal halide lamps are more suitable for colour sensitive applications.

Similarly, for application where colour sensitiveness is not so important like street lighting, sodium lamps are preferred since its efficacy is high in comparison with other lamps.

The next generation luminaries will be of LED (Light Emitting Diodes) type but these are still under development stage and presently, are being used in the following applications:

- Electrical panel indicating lamps
- Aviation lights
- Traffic signal lights
- Emergency light
- Fancy lights
- Portable lamp

A lot of improvement has taken place at reflector side also. Mirror optics types of reflectors are more popular at present in place of conventional acrylic reflectors.

5.5.4 Effect of Reduction in Supply Voltage on Energy Consumption

Power consumption of all types of lamps get reduced on reduction of supply voltage. But, voltage should be reduced to such a level that it should not cause flickering of the lamps. Transformers and reactors are available in the market which can reduce the voltage as per set value and at pre set time. Reduction in voltage affects the output of the light also to some extent. Therefore, the time for reduction in voltage is set in such a way that much light is not required during that period. This will result in saving in lighting energy.

5.5.5 Electronic Ballast and Low Loss Electromagnetic Choke

The conventional ballast consumes about 15 to 20 watts of power, while the power consumption of electronic ballast is approximately 1 watt only. The drawback of electronic ballast is that it injects harmonic in the system. It is also 1.5 to 2 times costlier from the conventional ballast. The other type of ballast is vacuum impregnated copper ballast. It consumes about 7 watt of power and is cheaper also in comparison with the electronic ballast.

5.5.6 Timers and Occupancy Sensors

Many microcontroller based, programmable times are available in the market. ON and OFF time can be set, so that the lights are switched off when not required. Infra red type occupancy sensors can also be used to switch off the lights. The other types of sensors are ultra sonic, which are not available in India.

Occupancy sensors are connected with microcontroller based system to switch off the lights when the occupancy is zero. Occupancy sensors are useful for the place which is occasionally occupied.

5.5.7 Fluorescent Tube Light

Fluorescent tube lights are of following types:

- T12: 40 watt, tube diameter is 12/8" (33.8mm)
- T8: 36 watt, tube diameter is 8/8" (26mm)
- T5: 14, 21, 28, 35 watts, tube diameter is 5/8" (16mm)

The diameter of T5 lamp is shorter than the diameter of T8 and T12 lamps. This increases efficiency of the lamp. T5 lamp is 50mm shorter than T8 and T12 in length and it can only work with electronic choke, therefore special arrangement is required for retrofitting the T5 lamp in the existing T12 or T8 lighting fixture.

The efficacy of 35 watt T5 lamp is 95 lm/watt, while T8 lamp is 89 lm/watt (with electronic choke). T5 lamp has a coating inside the glass that prevent the absorption of mercury into glass thus the use of mercury is drastically reduced.

5.5.8 CFL Lamps

Colour rendering index of CFL lamp is above 80 and lm/watt output ranges 55 to 67. Generally, CFL lamps are used in down-light luminaries in showrooms, galleries, hotel lobbies etc. CFL are much more efficient if compared with the normal GLS and halogen lamps which are used in down-light luminaries.

5.6 Energy Efficiency Building Code

Energy efficiency building code, has various provisions mentioned under various clauses related with building illumination system. These provisions are mandatory and applicable to the following area:

- Interior spaces of the building.
- Exterior building features including facades, illuminated roofs, architectural features, entrance, exits, loading docks and illuminated canopies.
- Exterior building ground lighting that is provided through the building's electrical service.

The following lightings are not covered in the mandatory requirement.

- Emergency lighting that is automatically off during normal building operation and is powered by battery, generator or any other alternate power source and
- Lighting in dwelling units.

5.6.1 Lighting Control

Automatic Lighting Shutoff

An area of larger than 500m^2 or 5,000ft^2 illuminated by interior lighting shall have automatic lighting control device. Within the building, if the office space of all the offices are less than 30m^2 or 300ft^2 and enclosed by walls up to ceiling height, conferences and meeting rooms, all shall be provided with occupancy sensor. In case of school all the classrooms shall be fitted with occupancy sensors. In addition to this, all the storage spaces shall also be fitted with occupancy sensors. For other locations following automatic control devices shall be provided.

- A programmed timer shall be provided for a area of not more than 2500 m^2 and not more than one floor.

Or

- Occupancy sensors should be provided to turn off the lights within 30 minutes of the occupant leaving the space. Light fixtures controlled by the occupancy sensors shall also have a manual facility like a manual on-off switch to turn off the lights even though the space is occupied.

The above is not applicable where, in exceptional cases the lighting system is designed for 24 hours use.

Space Controlled

General lighting provided in a space which is enclosed by ceiling height partition walls shall have at least one controlled device to control the lights. The control device shall be activated either manually or by sensor.

- Each control device shall control a maximum space of 250m^2 (2,500ft^2), when the total space is 1000m^2 (10,000ft^2) and a maximum of 1000m^2 (10,000ft^2) when the total space is greater than 1000m^2 (10,000ft^2).
- The control device shall be capable to override the shutoff of the lights, but for not more than two hours.

- The control device shall be located in such a way that it should be readily accessible.

It has also exception that it can be remotely located for safety and security reason. But it shall have a pilot light indication and clearly labelled to identify the control device.

Control in Daylight Area

Luminaries in day-lighted area which is greater than 25m^2 (250ft^2) shall be equipped with either a manual or automatic control device so that:

- the output of luminaries can be reduced at least by 50%,
- the luminaries located entirely within the daylight area could be controlled.

5.6.2 Exterior Lighting Control

All the exterior lighting shall be controlled by a photo sensor or an astronomical time switch which should be capable to switch off the lights when it is not required.

Additional Control

The following lighting application shall be equipped with control device to control the devices independently.

- Display or accent lighting where the area is greater than 300m^2 (3,000ft^2) shall have separate control device.
- Lighting used for display purpose shall have separate control devices if the area is greater than 300m^2 (3,000ft^2).
- Hotel and motel guest rooms and guest suits should have a master control device to control all the permanently installed luminaries and receptacles.
- Task lighting including permanently installed lighting under shelf and cabinet shall have control device integral to luminaries or a separately wall mounted device which is accessible and labelled clearly.
- Non-visual lighting such as plant growth and food warming shall be equipped with a separate control device.
- Lighting equipment that is for sale or for demonstration in lighting education shall be equipped with a separate control device accessible only to authorized personnel.

Exit Sign

The power consumption of internally illuminated exit sign shall not exceed 5 watt.

Exterior Building Ground Lighting

The luminaries provided for external building ground lighting and having a lamp of more than 100 watt, the lamp efficacy should be more than 60 lm/watt unless the luminary is controlled by a motion sensor.

Interior Lighting Power

The power consumption of interior lighting shall not exceed the interior lighting power allowance. The building code provides two methods for determination of lighting power allowance (watt) one is Building Area Method and another is Space Function Method.

Building Area Method

- Refer Table 5.5, where LDP (Lighting Power Density) for each type of building is given.
- Calculate the gross lighted floor area for each building.
- Multiply the area with LDP as specified for that particular area. This gives lighting power allowance for that particular area.
- Similarly, calculate the lighting power allowance for other area also and summation will give the total lighting power.

Table 5.5 Interior Lighting Power: Building Area Method

Building Area Type	*LPD (W/m²)*
Automotive Facility	9.7
Convention Centre	12.9
Court House	12.9
Dining: Bar Lounge/Leisure	14
Dining: Cafeteria/Fast Food	15.1
Dinning: Family	17.2
Dormitory	10.8
Exercise Centre	10.8
Gymnasium	11.8
Healthcare-clinic	10.8
Hospital/Healthcare	12.9
Hotel	10.8
Library	14
Manufacturing Facility	14
Motel	10.8
Motion Picture Theatre	12.9
Multifamily	7.5
Museum	11.8
Office	10.8
Parking Garage	3.2

Penitentiary	10.8
Performing Arts Theatre	17.2
Police/Fire Station	10.8
Post Office	11.8
Religious Building	14
Retail	16.1
School/University	16.1
Sports Arena	11.8
Town Hall	11.8
Transportation	10.8
Warehouse	8.6
Workshop	15.1

Space Function Method

- In this method for each space enclosed by partition walls, 80% or greater than the ceiling height, gross floor area is determined by measuring centre of the partition wall, including the floor area of balconies and other projections.
- The above area is multiplied by the LDP specified for that area in Table 5.6.
- Thus, the lighting wattage for each area is calculated and added to calculate the total wattage required for the lighting system.

Table 5.6 Interior Lighting Power: Space Function Method

Space Function	*LDP(W/m²)*
Office-enclosed	11.8
Office-open Plan	11.8
Conference/Meeting/Multipurpose	14.0
Classroom/Lecture/Training	15.1
Lobby	14.0
For Hotel	11.8
For Performing Art Theatre	35.5
For Motion Picture Theatre	11.8
Audience/Seating Area	9.7
For Gymnasium	4.3
For Exercise Centre	3.2
For Convention Centre	7.5
For Religious Buildings	18.3

For Sports Arena	4.3
For Performing Art Theatre	28.0
For Motion Picture Theatre	12.9
For Transportation	5.4
Atrium-first Three Floor	6.5
Atrium-each Additional Floor	2.2
Lounge/Recreation	12.9
For Hospital	8.6
Dining Area	9.7
For Hotel	14.0
For Motel	12.9
For Bar Lounge/Leisure Dining	15.1
For Family Dining	12.6
Food Preparation	12.9
Laboratory	15.1
Restrooms	9.7
Dressing/Locker/Fitting room	6.5
Corridor/Transition	5.4
For Hospital	10.8
For Manufacturing Facility	5.4
Stairs-active	6.5
Active Storage	8.6
For Hospital	9.7
Inactive Storage	3.2
Electrical/Mechanical	16.1
Workshop	20.5
Sleeping Quarters	3.2
Library	
Card File and Catalouging	11.8
Stacks	18.3
Reading Area	12.9
Hospital	
Emergency	29.1
Recovery	8.6
Nurse Station	10.8

Exam Treatment	16.1
Pharmacy	12.9
Patient Room	7.5
Operating Room	23.7
Nursery	6.5
Medical Supply	15.1
Physical Therapy	9.7
Radiology	4.3
Laundry-washing	6.5
Automotive Service Repair	7.5
Manufacturing	
Low Bay (< 8m ceiling)	12.9
High Bay (>8m ceiling)	18.3
Detailed Manufacturing	22.6
Equipment Room	12.9
Control Room	5.4
Hotel/Motel Guest Room	11.8
Dormitory-living Quarters	11.8
Museum	
General Exhibition	10.8
Restoration	18.3
Bank Office-banking Activity Area	16.1
Religious Building	
Worship-Pulpit, Choir	25.8
Fellowship Hall	9.7
Retail	
Sales Area	18.3
Mall Concourse	18.3
Sports Arena	
Ring Sports Area	29.1
Court Sports Area	14.8
Indoor Field Area	15.1
Sports Arena	
Ring Sports Area	2.7
Court Sports Area	2.3

Indoor Field Area	1.4
Warehouse	
Fine Material Storage	15.1
Medium/Bulky Material Storage	9.7
Parking: Garage Area	2.2
Transportation	
Airport-concourse	6.5
Air/Train/Bus-baggage Area	10.8
Terminal-ticket Counter	16.1

The following lighting equipments and applications shall not be considered while determining the interior lighting power allowance.

- Display or accent lighting is an essential element for the function performed in galleries, museums and monuments.
- Lighting that is integral to equipment or instrumentation and is installed by its manufacturer.
- Lighting that is specifically designed for medical or dental procedure and integral part of medical equipment.
- Lighting is an integral part of food processing and food warming.
- Lighting for plant growth and maintenance.
- Lighting in spaces specifically designed for use by visually impaired.
- Lighting in retail window display provided that displayed area is enclosed by ceiling height partitions
- Lighting for interior spaces that have been specifically designed as a registered interior historic landmark.
- Lighting that is an integral part of advertising or directional signage.
- Exit signs
- Lighting that is for sale or lighting education and demonstration system.
- Lighting for theatrical purpose including performance, stage and film or video production.
- Athletic playing area with permanent facilities for television broadcasting.

Installed Interior Lighting Power

Installed interior lighting power is the total power required for luminaries and its accessories like ballast, lamps, current regulators, control devices.

Luminary Wattage

It shall be determined as per the following:

- The wattage of incandescent luminaries with medium base sockets and not containing permanently installed ballasts shall be the maximum labelled wattage of the luminaries.
- The wattage of luminaries containing permanently installed ballasts shall be the operating input wattage of the specified lamp/ballast combination based on values from manufacturer catalogues or values from independents testing laboratory reports.
- The wattage of all other miscellaneous luminary types shall be specified wattage of the luminaries.
- The wattage of lighting track, plug-in bus-way, and flexible-lighting systems that allow the addition and/or relocation of luminaries without altering the wiring of the system shall be the larger of the specified wattage of the luminaries included in the system or 135 W/m. System with integral overload protection, such as fuses or circuit-breaker, shall be rated at 100% of the maximum rated load of the limiting device.

Exterior Lighting Power

As per this clause, the power requirement for exterior lighting for different location of the building, shall not exceed the specified limit as given in Table 5.7.

Table 5.7

Exterior Lighting Applications	*Power Limit*
Building Entrance (with canopy)	13W/M^2 (1.3 W/ft^2) of canopy area
Building Entrance	90W/lin metre (30W/lin feet) of door width
Building Exit	60 W/lin metre (20W/lin feet) of door width
Building Facades	2W/m^2 (0.2 W/ft^2) of vertical façade area

Lighting used for the following exterior applications is exempted when equipped with an independent control device.

- Specialized signal, directional, and marker lighting associated with transportation.
- Lighting used to highlight the features of public monuments and registered historic landmark structures and buildings.
- Lighting that is integral to advertising signage.
- Lighting that is specifically designed as required by a health or life safety requirement, ordinance, or regulation.

5.7 Conclusion

From the above, it can be concluded that lighting is an area where a lot of scope is available for improvement. Care should be taken at the building design stage, so that as far as possible natural daylight can be utilized at a maximum extent. Further, various techniques are elaborated for reduction of energy in lighting system. A detailed analysis is necessary to adopt a particular technique for a particular installation and sometimes a combination of techniques may be adopted.

5.8 Software

By using the software an estimate of scope of energy saving in lighting system can be made.

5.8.1 Objective

- To make a database of lighting fixtures for which energy audit is to be conducted.
- Estimation of energy saving.
- Calculation of payback period.
- To generate a print of the calculated values.

5.8.2 Procedure

- Click on "Electrical Equipment: Inventory" of main screen. A submenu will appear, click on "Lighting" (as shown by an arrow in Fig. 5.0). A screen as shown in Fig. 5.1 will appear.
- The procedure for entering the details for making database of lighting fixtures is same as mentioned in earlier chapters.
- Area-code is a specific code for area, where light fixtures are installed. It can be any alpha numerical or numerical figure.

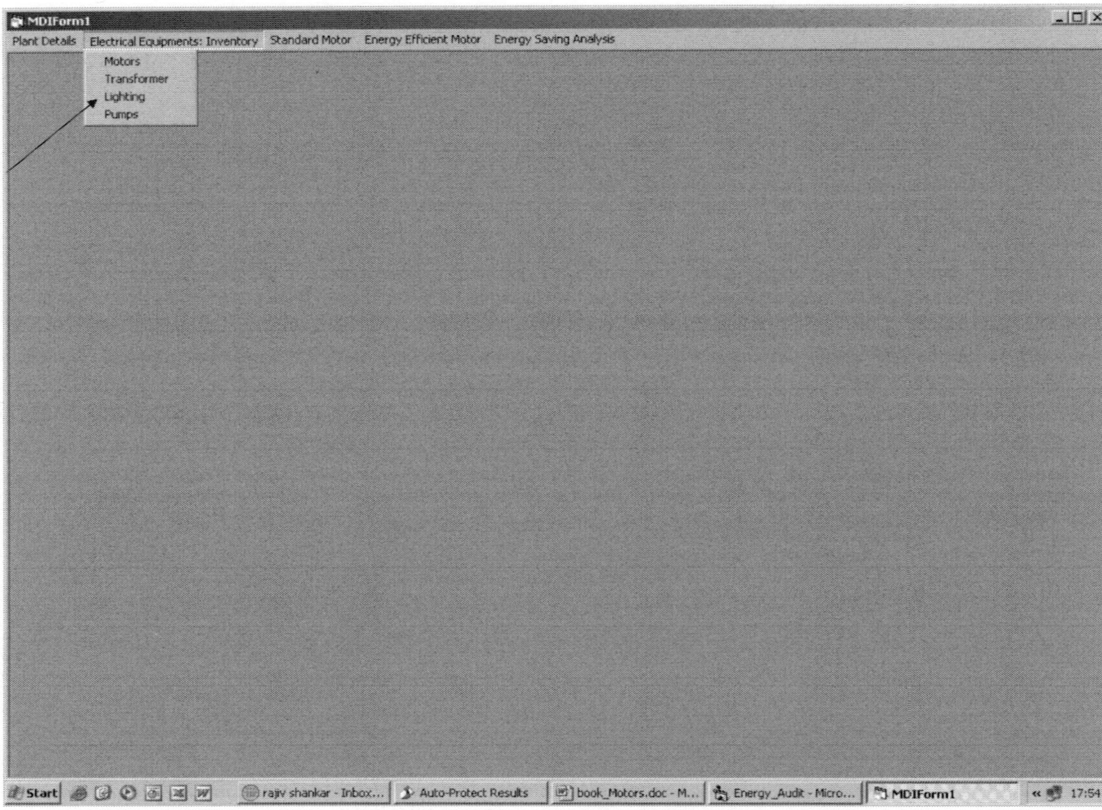

Fig. 5.0 Main screen

- After making a database of the lighting system, the next step is to perform analysis. Click on the "Energy Saving Analysis" of main screen. A submenu will open up. Click on "Lighting Analysis" of submenu as shown in Fig. 5.2. A screen as shown in Fig. 5.3 will appear. This screen is for analysis of the existing lighting system.
- To retrieve data from the database of lighting system, click on "Data Bank" as shown by an arrow in Fig. 5.3.

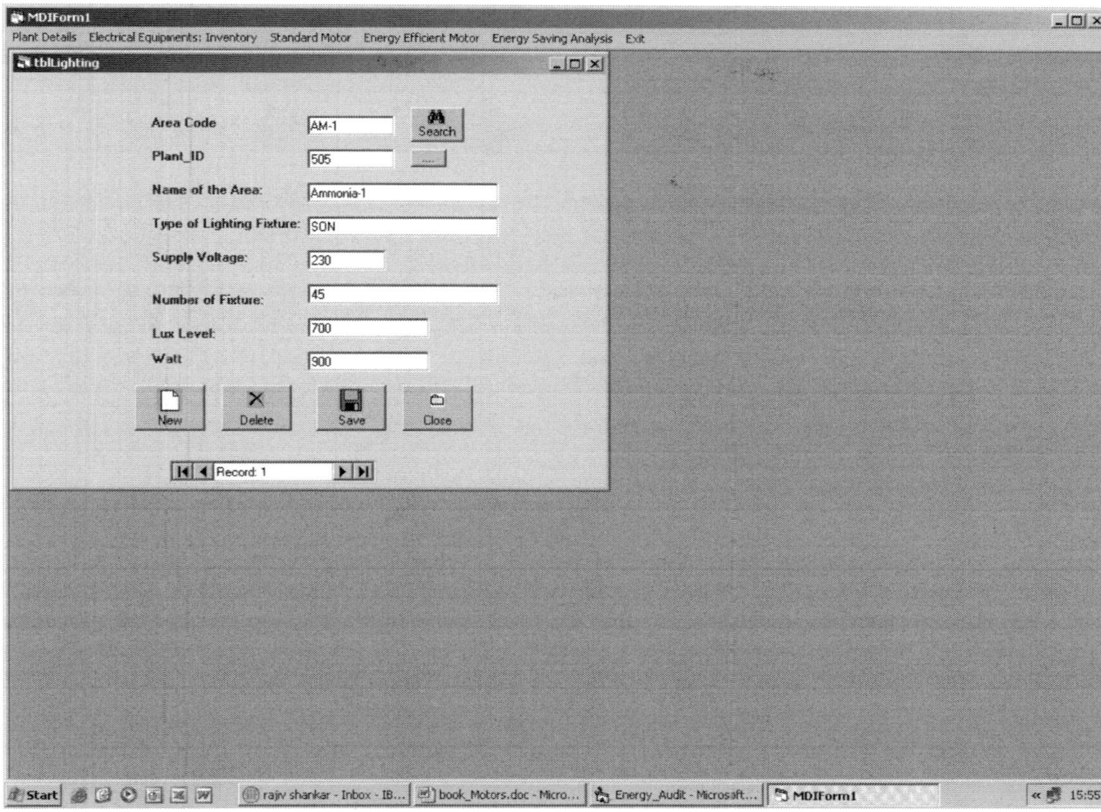

Fig. 5.1 Lighting details entry screen

- This clicking action will open a search screen as shown in Fig 5.4.
- Fill in the information like number of fixture, wattage and type of fixture in specified boxes to establish the search criteria. Click on "Search" button. The information as per the search criteria will be retrieved.
- Click on the "Area-code" to transfer the search results to the main lighting analysis screen.

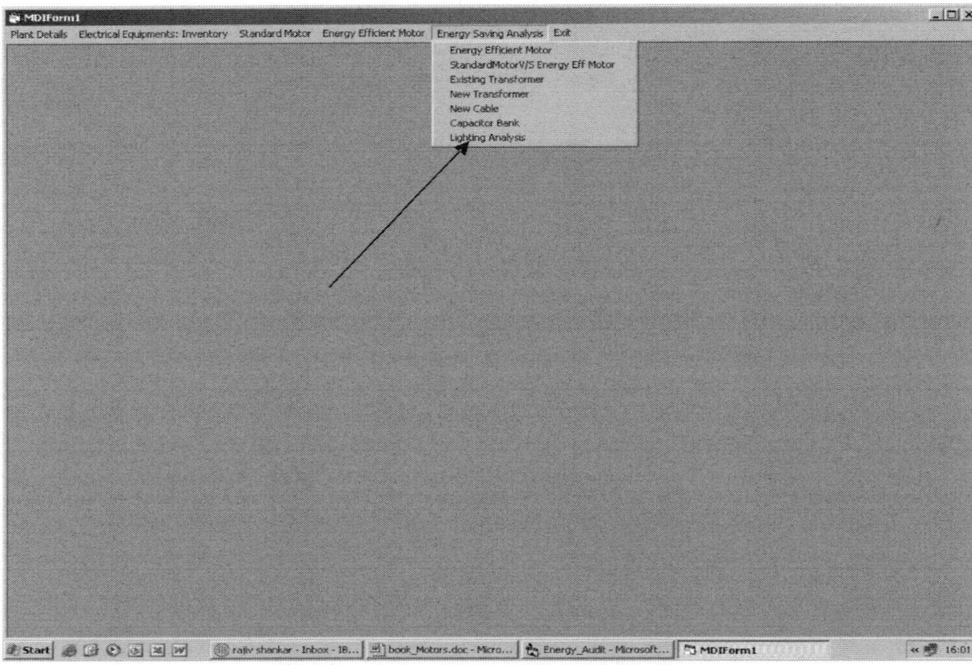

Fig. 5.2 Lighting analysis

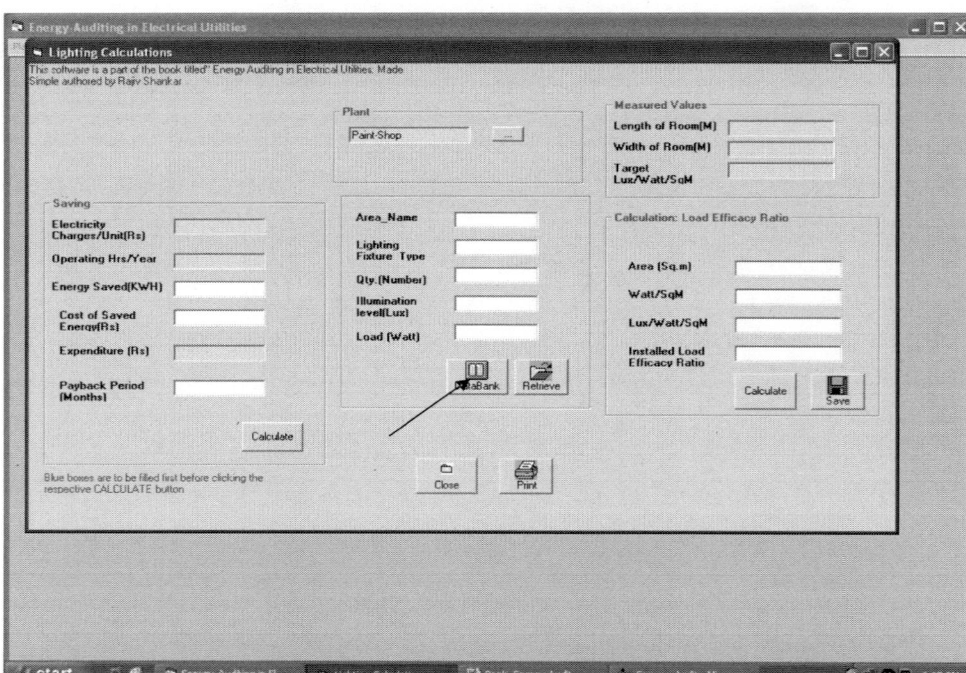

Fig. 5.3 Lighting analysis main screen

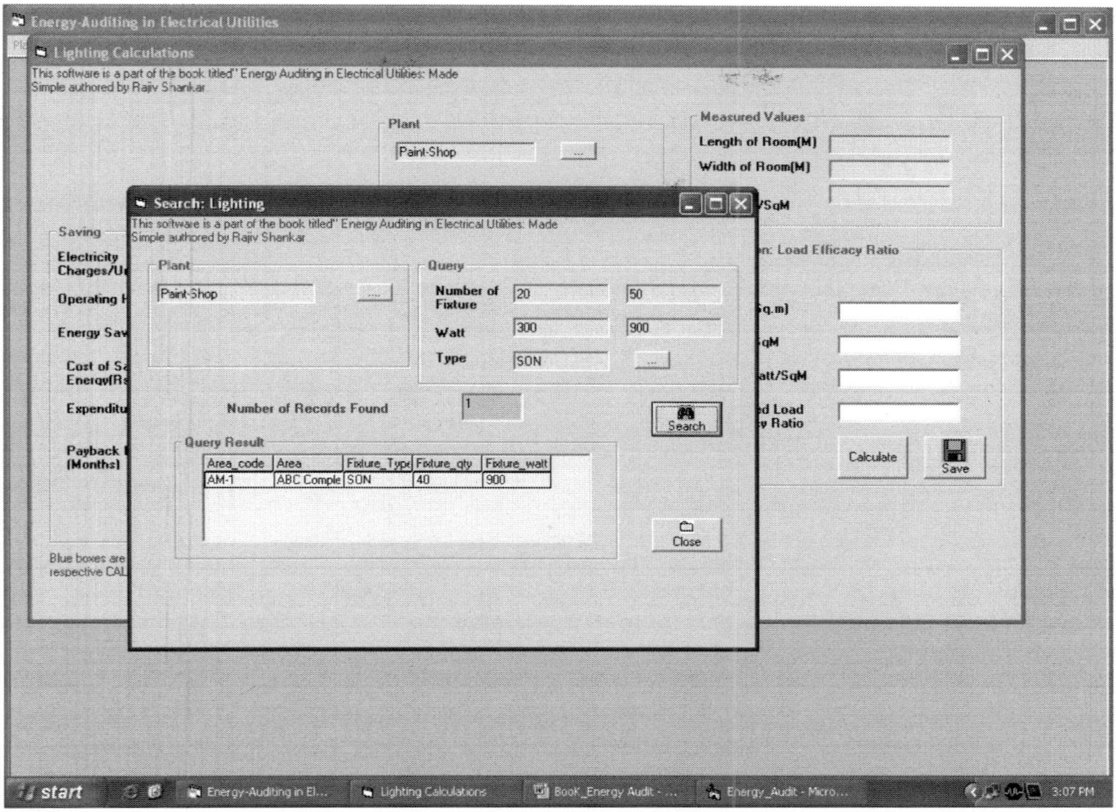

Fig. 5.4 Search for lighting data

- Click on "Retrieve" button to retrieve the data on the main screen as shown in Fig. 5.5
- Fill in the required information in the blank boxes like room dimensions, target lighting efficacy i.e. target lux/watt/area etc. as required. The information like electricity charges, operating hours, cost of material required for improvement of system etc. is also needed to be filled.
- Click on "Calculate" button to calculate the existing lighting efficacy and the installed load efficacy ratio as shown by a dotted arrow in Fig. 5.6.

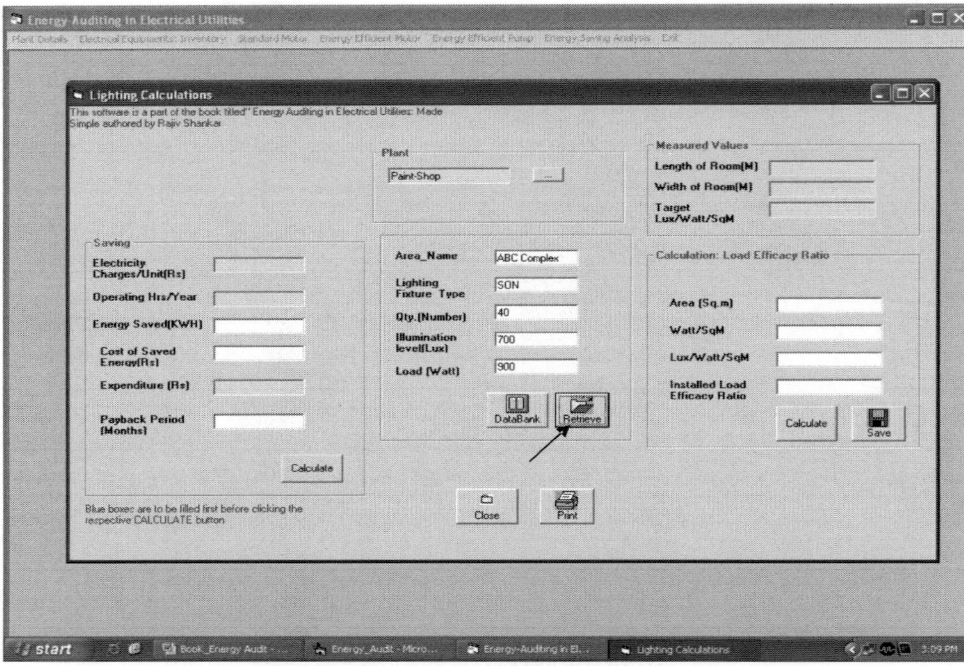

Fig. 5.5 Retrieval of data

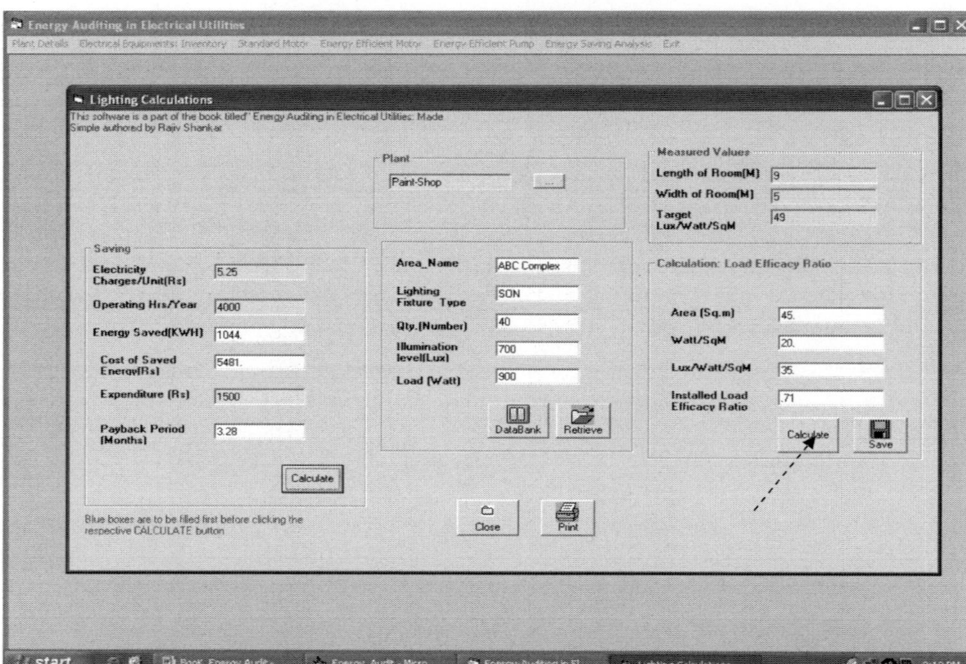

Fig. 5.6 Calculation of energy saved

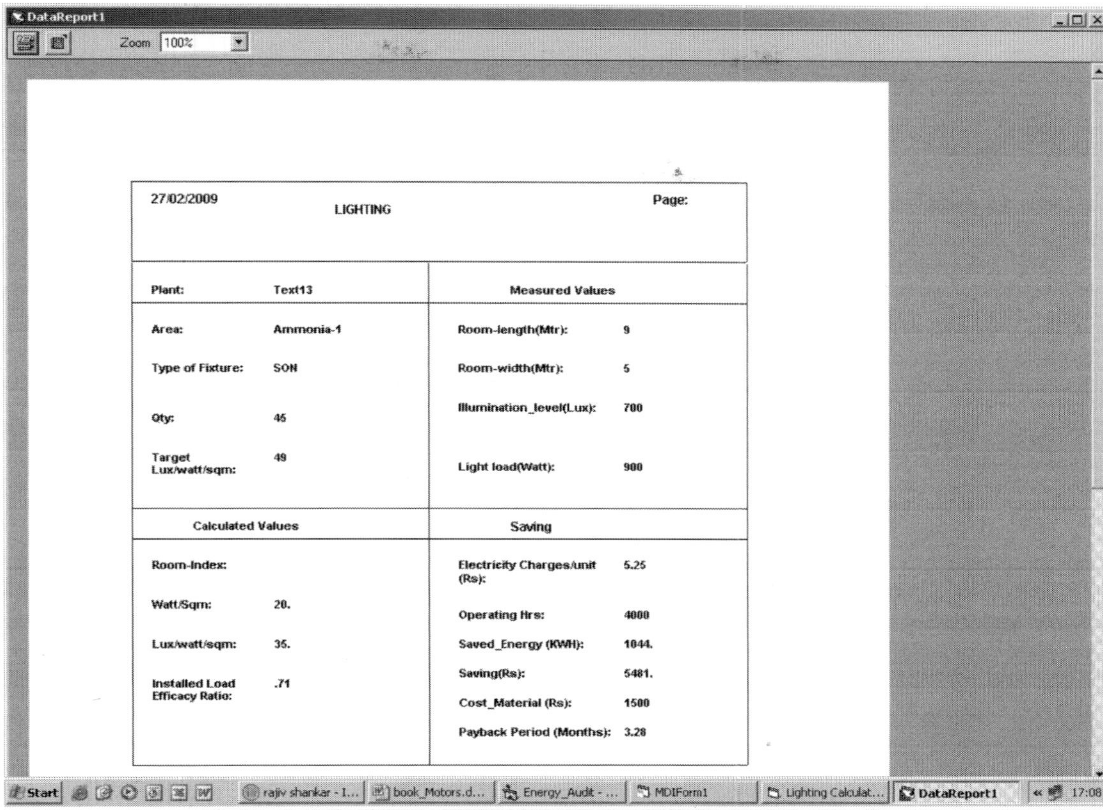

Fig. 5.7 Printout of analysis result

- Click on "Print" button to print the results of the calculations as shown in Fig. 5.7.

For calculating the target lighting efficacy, lux level required for that area is obtained from the IS:3646 and light power density LDP watt/m^2 can be taken from the standards like energy conservation building code. The ratio of lux and LDP will be the target lux/watt/m^2

To further clarify it, let's take an example.

Example 5.1

Following are the measured value.

Room length = 9 metre
Room width = 5 metre
Lux level = 700
Lighting load = 900 watt
Target lux/watt/m^2 = 49

Calculation:

Room area $= 45m^2$

Watt/m^2 $= 900/45 = 20$

Lux/watt/m^2 $= 700/20 = 35$

$$\text{Installed load efficacy ratio} = \frac{\text{Existing lux/watt/m}^2}{\text{Target lux/watt/m}^2}$$

$= 35/49$

$= 0.71$

Annual energy wastage $= (1 - \text{Installed load efficacy ratio}) \times \text{Watt} \times \text{Operating hours}$

$= (1 - 0.71) \times 900 \times 4000/1000$

$= 1044$KWH

Cost of energy which can be saved $= 1044 \times 5.25 =$ Rs5481

Expenditure made to save the energy = Rs1500

Simple payback period in months will be $= 1500 \times 12/5481$

$= 3.28$ months

If the installed load efficacy ratio is less than 0.7, then it is required to take necessary measures to improve the existing lighting installation by replacement it with more energy efficient lighting luminaries.

CHAPTER

6

Capacitor Bank

6.0 Introduction

In electrical distribution system, capacitor banks are installed to compensate the reactive power which is normally present in the distribution network due to reactive load. The reactive load puts burden on the distribution network of utility company. Therefore, utility company imposes penalty on the consumers if they draw reactive power beyond a certain limit. This penalty increases the cost of electricity purchased from utility company.

The additional benefits of reduction in reactive power in the network are reduction in I^2R losses, improvement in system efficiency, release of additional capacity, reduction in size of electrical equipment for delivery of same amount of power and improvement in voltage regulation.

6.1 Reactive Power

Any electrical system will always have some reactive load. Some of the examples of reactive loads are:

- Motors
- Transformers
- Fluorescent lamps
- Rectifiers
- Furnace
- Various types of electronic equipment
- Computers

The current drawn by the electrical equipment has two components, Reactive and Active Component. The active component is responsible for doing useful work like producing power at a motor shaft. The reactive component, which is also called watt-less component, is responsible for producing necessary flux across the air gap of an induction motor or in the core of a transformer. For transmission of electrical energy across the air gap between stator of motor and rotor or through the core of the transformer, building up flux for magnetic field is essentially required. Without this magnetic field, electrical energy can not flow across the air gap of a motor or in the core of a transformer. To generate this magnetic field, magnetizing current flows in the system. The total current as read in the ammeter is the vector sum of these two components. In an inductive load the reactive component lags at 90° from the active component.

This lagging component is responsible for power factor. The power factor is affected by the reactive component. If the magnetizing current (reactive component) is more in comparison to the active component, the power factor will be poor as in a lightly loaded induction motor.

Now, the question arises that how the poor power factor affects the utility and why the utility companies fix a power factor limit and imposes penalty on the consumers, if the power factor falls below that limit?

More reactive current in comparison of active current for a load will result in poor power factor. As mentioned earlier, since reactive current is also a part of the total current supplied by the utility, The equipments of the utility like transmission lines, transformers, switch-gears etc. will have to carry more current resulting in additional load on the equipments of utility company. In another way, we can say that for supplying the same useful power to a load of poor power factor more capacity of the equipment is utilized in comparison with a load of good power factor.

6.2 Power Factor

Power factor is the ratio of active power to the apparent power. For example, an electrical panel consists of an ammeter, voltmeter and wattmeter are provided. The ammeter is showing 200 ampere, voltmeter is showing 415 volts and wattmeter is showing 120KW. Then apparent power will be $(1.732 \times 200 \times 415)/1000 = 143.76$ KVA and the power factor will be $120/143.76 = 0.83$. Trigonometrically, the power factor is $KW/KVA = \cos \phi$. Generally, any power factor below 0.9 is called Poor Power Factor.

6.3 Remedy of Poor Power Factor

The remedy is to install a capacitor bank or sometimes synchronous motors to improve the power factor. The reactive power required is supplied locally by the capacitor bank and in this way utility is relived from supplying the reactive power to the load. Before, installation of capacitor bank we should look into the following points:

- Limit of the power factor imposed by the utility company.
- Existing power factor and capacity of system and also the existing load on the system.
- Cost of adding additional capacitor banks to improve the power factor beyond

the limit as imposed by the utility company, release of additional capacity and also its cost.

- Payback period of installation of capacitor bank.
- Rebate provided by utility company if the power factor improves beyond the specified limit.

If the system has enough spare capacity and additional capacity is not required, the power factor should be kept as per the limit imposed by the utility company, which is normally 0.95, since 0.95 is the most economical power factor. Addition of more capacitor banks may not be economical if additional capacity of the system is not required. However, if the utility company provides rebate on the improved power factor beyond the limit specified by them, then it will be beneficial to provide capacitor banks to avail the benefit of rebate.

6.4 Location of Capacitor Bank

The best location for installation of capacitor banks are as near as possible to the load. Sometimes, capacitor banks are installed to act as group correction i.e. it is connected with the main bus bar and the main bus has many feeders, since it is not always economical to install the capacitor for individual feeders. For example, if the motors of different rating are installed, many capacitors of different ratings will be needed resulting in increase in cost. Group correction is also economical in cases when different feeders have different load at different time. Because in such cases automatic power factor sensing relay continuously keep on sensing the power factor of the load and keep on switching on and off the capacitor banks to adjust the power factor as per the pre set value.

The another important point, which should be taken care of, is that the improvement takes place only from point of application towards the source of power and not in the opposite direction. Therefore, it is always advantageous to place the capacitor bank near the load that reduces the losses in the circuit between the load and the feeder.

6.5 Economics of Power Factor Improvement

Electricity tariff mainly consists of two parts, one is Demand Charges and another is Unit Charges. Demand charges are KVA charges and it depends on power factor. If power factor is low, the KVA will be more and vice versa. In this way the electricity bill for poor power factor will be more due to more KVA demand. The unit(KWH) charges will not be affected. By installing capacitor bank, the intake of reactive power from the utility grid will get reduced, since it is now supplied by the capacitor banks provided locally. This will reduce the KVA demand and subsequently KVA demand charges will also get reduced.

We take an example to illustrate the above aspect.

Example 6.1

As shown in the single line diagram, supply is taken at 11KV and stepped down to 433 volt through 11KV/433V, 1000KVA, Delta-star connected step down transformer. After stepping down it feeds to Bus-B of power control centre. Bus-B consists of three feeders.

- Feeder 1 is connected to a motor control centre having a load of 500KW at 0.85 power factor. The power supply to motor control centre is fed through a 630 sq.mm, single core, PVC insulated, aluminium conductor cables. Three numbers of such cables are provided in each phase and two numbers in neutral. The approximate distance between MCC feeder of Bus-B and the incomer of Bus-C (bus of motor control centre) is 400 metres.

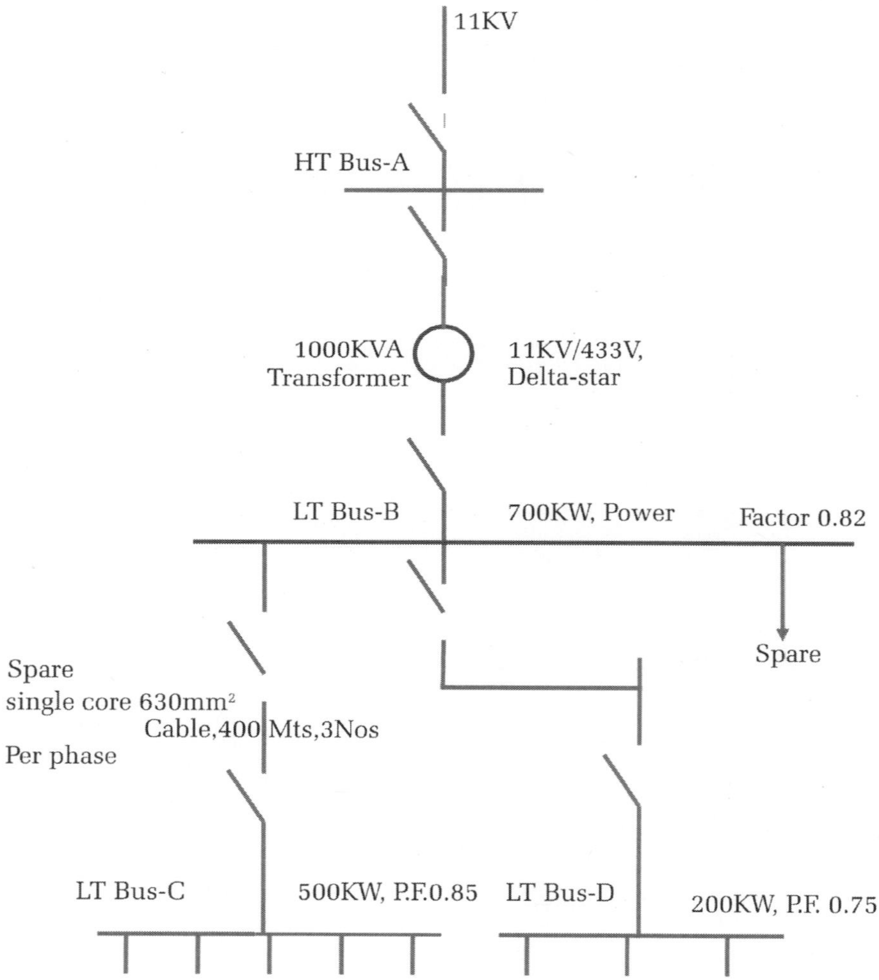

Fig. 6.0

- Feeder 2 is connected to a miscellaneous load consisting of lighting load, lifts load, window air conditioner, computers etc. The total load is 200KW at 0.75 power factor. It is fed through two number, three and a half core, 240 sq.mm, aluminium conductor, PVC insulated cables. The distance is approximately 40 metres.
- The third feeder is spare. The utility has imposed a condition that the power factor should not be less than 0.95. The demand charges are Rs250.00 per KVA. Unit charges are Rs4.60 per unit.

  We would do analysis for working out the KVAR requirement and cost of the capacitor banks for improving the power factor and also for the best location of the bank.

Analysis

Case 1

Capacitor Banks are provided at Bus-B (group compensation).

The total load on Bus-B is 700KW and combined power factor will be 0.82. For improving power factor from 0.82 to 0.95 the KVAR required can be calculated from the following formula:

$$KW(\tan \phi 1 - \tan \phi 2),$$

Where $\cos \phi 1$ is existing power factor and $\cos \phi 2$ is the improved power factor. Alternatively, standard tables (refer Table 6.1) are also available to determine the KW multiplying factor for determining the required KVAR for required power factor correction.

Table 6.1

Existing Power Factor						*Impr-oved*	*Power*	*Factor*					
	0.8	0.85	0.90	0.91	0.92	0.93	0.94	0.95	0.96	0.97	0.98	0.99	1.00
0.40	1.537	1.668	1.805	1.832	1.861	1.895	1.924	1.959	1.998	2.037	2.085	2.146	2.228
0.41	1.474	1.605	1.742	1.769	1.798	1.831	1.860	1.896	1.935	1.973	2.021	2.082	2.225
0.42	1.413	1.544	1.681	1.709	1.738	1.771	1.800	1.836	1.874	1.913	1.961	2.022	2.164
0.43	1.356	1.487	1.624	1.651	1.680	1.713	1.742	1.778	1.816	1.855	1.903	1.964	2.107
0.44	1.290	1.421	1.558	1.585	1.614	1.647	1.677	1.712	1.751	1.790	1.837	1.899	2.041
0.45	1.230	1.360	1.501	1.532	1.561	1.592	1.626	1.659	1.695	1.737	1.784	1.846	1.988
0.46	1.179	1.309	1.446	1.473	1.502	1.533	1.567	1.600	1.636	1.677	1.725	1.786	1.929
0.47	1.130	1.260	1.397	1.425	1.454	1.485	1.519	1.552	1.588	1.629	1.677	1.758	1.881
0.48	1.076	1.206	1.343	1.370	1.400	1.430	1.464	1.497	1.534	1.575	1.623	1.684	1.826
0.49	1.030	1.160	1.297	1.326	1.355	1.386	1.420	1.453	1.489	1.530	1.578	1.639	1.782
0.50	0.982	1.112	1.248	1.276	1.303	1.337	1.369	1.403	1.441	1.481	1.529	1.590	1.732

0.51	0.936	1.066	1.202	1.230	1.257	1.291	1.323	1.357	1.395	1.435	1.483	1.544	1.687
0.52	0.894	1.024	1.160	1.188	1.215	1.149	1.281	1.315	1.353	1.393	1.441	1.502	1.643
0.53	0.850	0.980	1.116	1.144	1.171	1.205	1.237	1.271	1.309	1.349	1.397	1.458	1.600
0.54	0.809	0.939	1.075	1.103	1.130	1.164	1.196	1.230	1.268	1.308	1.356	1.417	1.559
0.55	0.769	0.899	1.035	1.063	1.090	1.124	1.136	1.190	1.228	1.268	1.316	1.377	1.519
0.56	0.730	0.860	0.996	1.024	1.051	1.085	1.117	1.151	1.189	1.229	1.277	1.338	1.480
0.57	0.692	0.822	0.958	0.986	1.013	1.047	1.079	1.113	1.151	1.919	1.239	1.300	1.442
0.58	0.655	0.785	0.921	0.949	0.976	1.010	1.042	1.076	1.114	1.154	1.202	1.263	1.405
0.59	0.618	0.748	0.884	0.912	0.939	0.973	1.005	1.039	1.077	1.117	1.165	1.226	1.369
0.60	0.584	0.714	0.849	0.878	0.905	0.939	0.971	1.005	1.043	1.083	1.313	1.192	1.333
0.61	0.549	0.679	0.815	0.843	0.870	0.904	0.936	0.970	1.008	1.048	1.096	1.157	1.299
0.62	0.515	0.645	0.781	0.809	0.836	0.870	0.902	0.936	0.974	1.014	1.062	1.123	1.266
0.63	0.483	0.613	0.749	0.777	0.804	0.838	0.870	0.904	0.942	0.982	1.030	1.091	1.233
0.64	0.450	0.580	0.716	0.744	0.771	0.805	0.837	0.871	0.909	0.949	0.997	1.058	1.201
0.65	0.419	0.549	0.685	0.713	0.740	0.774	0.806	0.840	0.878	0.918	0.966	1.027	1.169
0.66	0.388	0.518	0.654	0.682	0.709	0.743	0.775	0.809	0.847	0.887	0.935	0.996	1.138
0.67	0.358	0.488	0.624	0.652	0.679	0.713	0.745	0.779	0.817	0.857	0.905	0.966	1.108
0.68	0.329	0.459	0.595	0.623	0.650	0.684	0.716	0.750	0.788	0.828	0.876	0.937	1.078
0.69	0.299	0.429	0.565	0.593	0.620	0.654	0.686	0.720	0.758	0.798	0.840	0.907	1.049
0.70	0.270	0.400	0.536	0.564	0.591	0.625	0.657	0.691	0.729	0.769	0.811	0.878	1.020
0.71	0.242	0.372	0.508	0.563	0.563	0.597	0.629	0.663	0.701	0.741	0.785	0.850	0.992
0.72	0.213	0.343	0.479	0.507	0.534	0.568	0.600	0.634	0.672	0.712	0.754	0.821	0.964
0.73	0.186	0.316	0.452	0.480	0.507	0.541	0.573	0.607	0.648	0.685	0.727	0.794	0.936
0.74	0.159	0.289	0.425	0.453	0.480	0.514	0.546	0.580	0.618	0.658	0.700	0.766	0.909
0.75	0.132	0.262	0.398	0.426	0.453	0.487	0.519	0.553	0.591	0.631	0.673	0.740	0.882
0.76	0.105	0.235	0.371	0.399	0.426	0.460	0.492	0.526	0.564	0.604	0.652	0.713	0.855
0.77	0.079	0.209	0.345	0.373	0.400	0.434	0.466	0.500	0.538	0.578	0.620	0.685	0.829
0.78	0.053	0.183	0.319	0.347	0.374	0.408	0.440	0.474	0.512	0.552	0.594	0.659	0.802
0.79	0.026	0.156	0.292	0.320	0.347	0.381	0.413	0.447	0.485	0.525	0.567	0.633	0.776
0.80		0.130	0.266	0.294	0.321	0.355	0.387	0.421	0.459	0.499	0.541	0.609	0.750
0.81		0.104	0.240	0.268	0.295	0.329	0.361	0.395	0.433	0.473	0.515	0.581	0.724
0.82		0.078	0.214	0.242	0.269	0.303	0.335	0.369	0.407	0.447	0.489	0.555	0.698
0.83		0.052	0.188	0.216	0.243	0.277	0.309	0.343	0.381	0.421	0.463	0.529	0.672
0.84		0.026	0.162	0.190	0.217	0.251	0.283	0.317	0.355	0.395	0.437	0.503	0.646
0.85			0.136	0.164	0.191	0.225	0.257	0.291	0.329	0.369	0.417	0.477	0.620
0.86			0.109	0.140	0.167	0.198	0.230	0.264	0.301	0.343	0.390	0.450	0.593
0.87			0.083	0.114	0.141	0.172	0.204	0.238	0.275	0.317	0.364	0.424	0.567
0.88			0.054	0.085	0.112	0.143	0.175	0.209	0.246	0.288	0.335	0.397	0.540
0.89			0.028	0.059	0.086	0.117	0.149	0.183	0.230	0.262	0.309	0.369	0.512
0.90				0.031	0.058	0.089	0.121	0.155	0.192	0.234	0.281	0.341	0.484
0.91					0.027	0.058	0.090	0.124	0.161	0.203	0.250	0.313	0.456
0.92						0.027	0.063	0.097	0.134	0.176	0.223	0.283	0.426
0.93							0.032	0.066	0.103	0.145	0.192	0.252	0.395
0.94								0.034	0.071	0.113	0.160	0.220	0.363
0.95									0.037	0.079	0.126	0.186	0.329
0.96										0.042	0.089	0.149	0.292
0.97											0.047	0.108	0.251
0.98												0.060	0.203
0.99													0.143

The multiplying factor for improving power factor from 0.82 to 0.95 is 0.369. Therefore, the KVAR required is 0.369 × 700 = **258.3**

Assuming the cost of supply and installation of capacitor bank per KVAR is Rs980.00, then the total cost of installation of capacitor banks will be Rs(258.3 × 980) = **Rs2,53,134.00**

Payback Period

KVA at 0.82 power factor will be 700/0.82 = 853.66

KVA charges

853.66 × 250 = Rs2,13,415.00

KVA at 0.95 power factor will be 700/0.95 = 736.84

KVA charges

736.84 × 250 = Rs1,84,210.00

Difference in demand charges Rs(2,13,415 – 1,84,210) = Rs29,205.00

i.e. saving per month = **Rs29,205.00**

Payback period will be 2,53,134/29,205 = **8.66 months** say approximately eight and a half month.

Additional capacity released = 853.66 – 736.84 = **116.82KVA**

Similarly, we can do the calculation for 0.99 power factor improvement.

The KVAR required will be **388.5** and charges for capacitor bank will be **Rs3,80,730.00.** The saving will be Rs36,650.00 per month and the payback period will be **10.38 months.**

Additional capacity released = 853.66 – 707 = 146.66KVA

Cost of per KVA capacity released when power factor if improved to 0.95 will be 2,53,134/116.84 = **Rs2,166.50**

Cost of per KVA capacity released when power factor is improved to 0.99 will be Rs3,80,730 /146.66 = **Rs2,596.00**

Case 2

Capacitor banks are provided at Bus-C and Bus-D (individual correction).

Bus-C

The multiplying factor for improving power factor from 0.85 to 0.95 is 0.291.

Therefore, the KVAR required is 0.291 × 500 = **145.5**

The total cost of installation of capacitor banks will be Rs(145.5 × 980) = **Rs1,42,590.00**

Saving in demand charges per month = 61.68 × 250 = **Rs15,421.00**

Bus-D

The multiplying factor for improving power factor from 0.75 to 0.95 is 0.553.

Therefore, the KVAR required is 0.553 × 200 = **110.6**

The total cost of installation of capacitor banks will be Rs(110.6 × 980) = **Rs1,08,388.00**

Saving in demand charges per month = 56.17 × 250 = Rs14,036.00

Total cost = Rs(1,42,590 + 1,08,388) = **Rs2,50,978.00**

Total saving in demand charges per month =29,457.00

We can summarize the above analysis as following:

- The main benefit in power factor improvement is reduction in electricity bill due to reduction in demand charges.
- If the length of feeding cable is more then we will get the additional benefit of reduction in I^2R losses, and saving in unit charges if the capacitor bank is installed at the load end as seen in the above case study in case of individual correction.
- It is not always economical that the power factor should be near to unity as we have seen that the payback period for improving power factor to 0.99 is more than the power factor of 0.95. And also per KVA cost of released capacity is more if power factor is improved to 0.99 instead of 0.95.

6.6 Energy Conservation Building Code

Building code states that all electricity supplies exceeding 100A, 3phase shall maintain their power factor between 0.98 lag and unity at the point of connection.

Further, building code also specifies about the metering and monitor of maximum demand, power factor, consumption of unit etc.

6.6.1 Metering and Monitoring

- Building whose maximum demand is greater than 250KVA shall have the electrical distribution system with their energy consumption being check-metred.
- Services exceeding 1000KVA shall have permanently installed electrical metering to record demand (KVA), energy (KWH) and total power factor. The metering shall also display current in each phase and the neutral, voltages between phases and between phase and neutral and total harmonic distortion as a percentage of total current.
- Services not exceeding 1000KVA but over 65KVA shall have permanently installed electric metering to record demand (KW), energy (KWH), and total power factor.
- Services not exceeding 65KVA shall have permanently installed electrical metering to record energy (KWH), KW and power factor.

6.6.2 Thermal Monitoring of Main Distribution System

All the power junction boxes and main power distribution board and cable termination points shall be provided with temperature monitoring mechanism comprising of sensors in the enclosed chambers and properly visible temperature indicators outside. Records of temperature during commissioning and subsequently on daily basis shall be maintained.

6.6.3 Power Distribution System Losses

The power cable shall be adequately sized as to maintained the distribution losses not to exceed 1% of the total power usage. Record of design calculation for the losses shall be maintained.

6.7 Conclusion

The main advantage of power factor improvement is reduction in KVA demand, which results in reduced power cost. In addition, the losses in the system also get reduced and a saving in energy is realized. But as compared to saving which is obtained in KVA charges, it is very small and installation of capacitor bank just for saving due to reduction in losses is not justified. It should always be looked as additional benefit.

Before installation of capacitor bank, measurement of harmonic in the system shall be made and if harmonics are more than the permissible limit, harmonic filter shall be provided since presence of harmonics adversely affects the performance of capacitor bank.

6.8 Software

6.8.1 Objective

- Calculation of KVAR required for improving in the existing power factor to its target value.
- Calculation of reduction in losses from improvement in the power factor.
- To calculate saving in electricity bill and payback period.
- To print the calculated results.

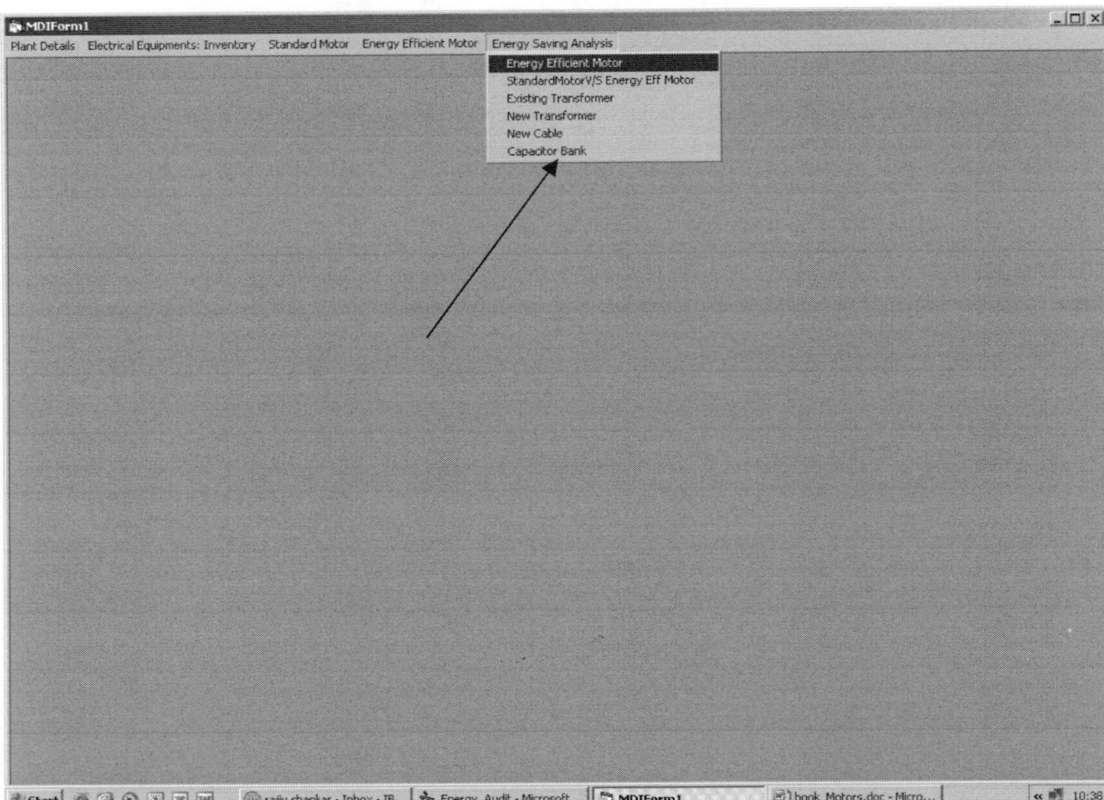

Fig. 6.1 Main screen

6.8.2 Procedure

- Click on the "Energy Saving Analysis" of main menu and then click on "Capacitor Bank" of submenu as shown by an arrow in Fig. 6.1. A screen as shown in Fig. 6.2 will appear.

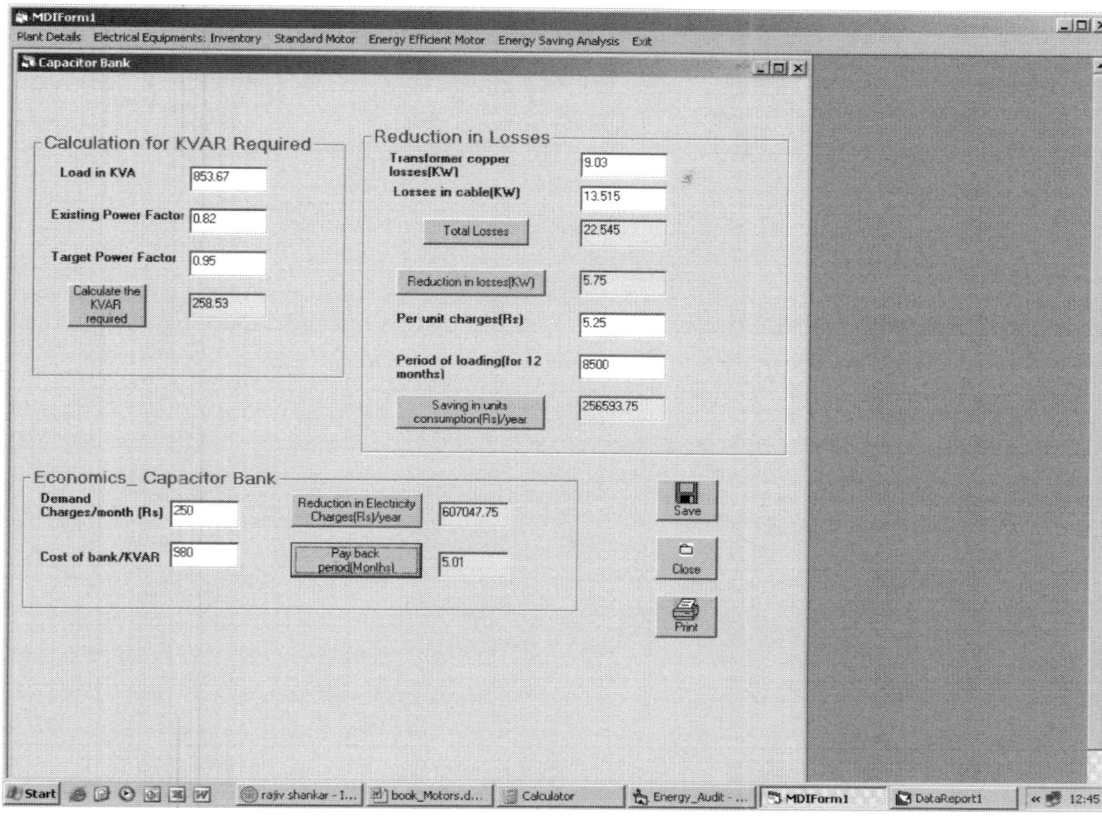

Fig. 6.2 Calculation for KVAR requirement

- Fill in the load in KVA and the existing and target power factor. To calculate the KVAR, click on the "Calculate KVAR Required" button.
- Similarly, reduction in losses can be calculated. Fill in I^2R losses of transformer and cables in the respective fields then click on "Total Losses" and "Reduction in Losses" buttons, total losses as well as reduction in losses due to improvement in power factor is calculated.
- Click on "Saving in Units Consumed" button to calculate saving in electricity consumption due to reduction of losses.
- Fill in demand charges per KVA and the installation charges per KVAR and click on "Reduction in Electricity Charges Per Year" button and "Payback Period" button to calculate the reduction in yearly bill and payback period for installation of capacitor bank respectively.

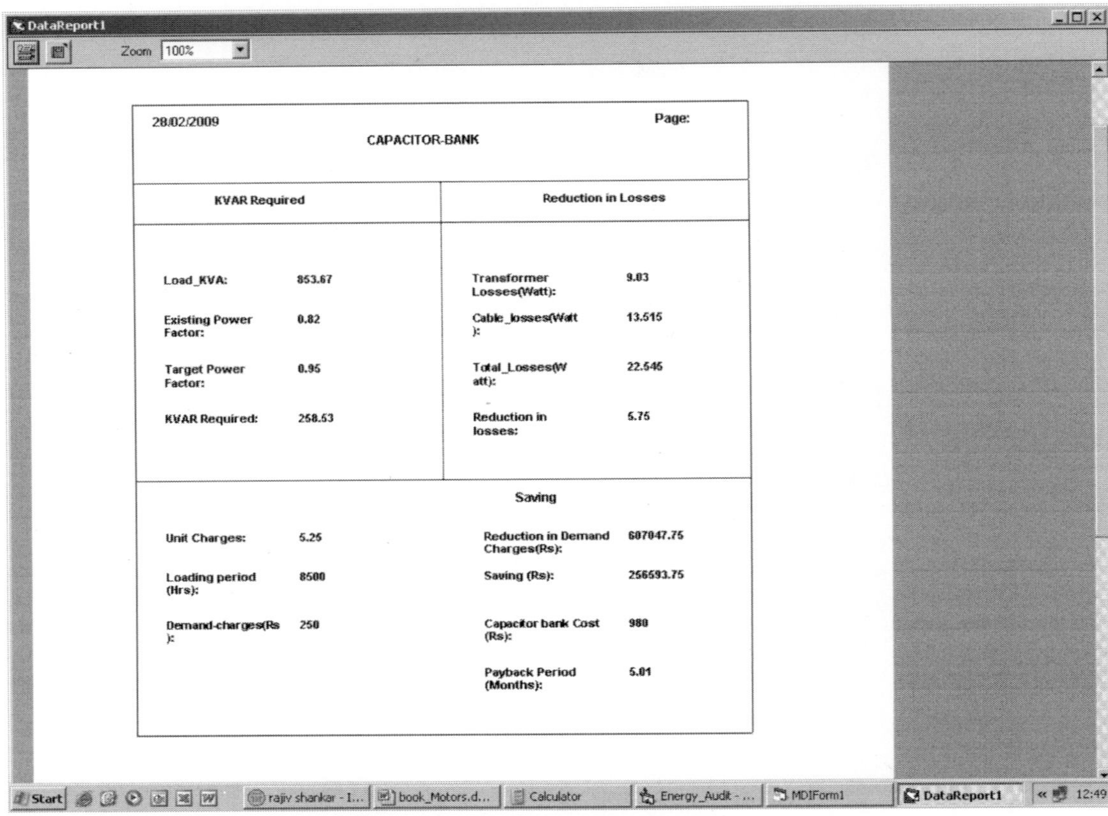
28/02/2009 Page:

CAPACITOR-BANK

KVAR Required		Reduction in Losses	
Load_KVA:	853.67	Transformer Losses(Watt):	9.03
Existing Power Factor:	0.82	Cable_losses(Watt):	13.515
Target Power Factor:	0.95	Total_Losses(Watt):	22.545
KVAR Required:	258.53	Reduction in losses:	5.75

Saving

Unit Charges:	5.25	Reduction in Demand Charges(Rs):	607047.75
Loading period (Hrs):	8500	Saving (Rs):	256593.75
Demand-charges(Rs):	250	Capacitor bank Cost (Rs):	980
		Payback Period (Months):	5.01

Fig. 6.3 Printout of the analysis result

- Click on "Save" button and then "Print" button to print the calculated results as shown in Fig. 6.3.

Example 6.2

To illustrate the above let's take an example.

We have the following parameters:

Current for 500KW, three phase, load at 0.82 power factor = 848.32 amp

Resistance of 630 sq.mm cable =0.047 ohm/kilometer

Resistance of 400 metres length =.0188 ohm

Since three cables are provided in parallel, the total resistance of cable in one phase will be one-third i.e. 0.00626 ohm.

I^2R losses in one phase $= (848.32/3)^2 \times 0.00626$

$=500.1$ watt

The losses for three phases $= 3 \times 500.1 = 1.5$KW

Example 6.3

Calculation for Losses in Transformer

Rating of transformer = 1000KVA
Loading of transformer = 853.65KVA
Power factor = 0.82
Copper losses at rated load = 12.5KW
Load factor = 853.65/1000 = 0.85
Copper losses of transformer = $0.85^2 \times 12.5$ = 9.03KW
Total losses of transformer and cable
= 1.5 + 9.03 = 10.53KW
Reduction in losses due to improvement in power factor
KW × (1/pf1 – 1/pf2)
pf_1 = Existing power factor
pf_2 = Improved power factor
KW = Active Power
= 10.53 × (1/0.82 – 1/0.95)
= 0.166KW
Unit (KWH) charges = Rs5.25 per unit
Period of loading of transformer and cable per year = 8500 hours
Saving in unit charges = 8500 × 5.25 × 0.166
= Rs7424
Reduction in demand charges per month = 29205 (refer Example 6.1)
Yearly reduction in demand charges = 12 × 29205 = Rs350460
Total saving in electricity bill = Rs7424 + Rs350460
= Rs357884.00
Cost of installation of capacitor bank = Rs253134 (refer Example 6.1)
Payback period = 253134 × 12/357884 = 8.4 months

CHAPTER

7

Compressed Air System

7.0 Introduction

In an industrial plant, particularly in a process plant, we will find many types of compressors like reciprocating, centrifugal, screw type etc. Some of the compressors are used for compressed air system; compressed air is required for opening and closing of pneumatic valves, instrumentation system, pneumatic tools etc.

Compressed air system can be divided into two parts. One part is Supply End and another is User End. Supply end consists of air compressor, air filters, dryers, heaters, storage system etc. User end consists of equipments where compressed air is used. The objective is to manage both the ends in such a way that clean and dry air is available at the required pressure and air wastage through leakages or misuse is minimized. Compressed air system can be an area where sufficient scope is available for energy saving.

Let's have a look into various types of air compressor.

7.1 Types of Air Compressor

Air compressor can be divided into two types.

- Positive displacement
- Dynamic

In a positive displacement, air or gas is first trapped in a compression chamber and then its volume is reduced mechanically, so that the pressure increases, while in dynamic compressor the impellers rotate at a very high speed and impart velocity energy to the air or gas. This velocity energy is changed into pressure by the impellers and the diffusers. Reciprocating and rotary compressors are the positive displacement type of air compressors while centrifugal and axial are the dynamic type of air compressors.

7.1.1 Reciprocating Compressor

Basically reciprocating compressor is available in four configurations.

- Horizontal
- Vertical
- Horizon balance opposed
- Tandem

The reciprocating compressor works on the principle of bicycle pump and consists of piston, cylinder, inlet valve, discharge valve etc. when compression is accomplished using only one side of the piston it's called Single Acting and when both side of the piston are used then it is called Double Acting Type.

The reciprocating type compressor may be of single stage or multistage type. In a single stage type compressor, the compression takes place in one cylinder, while in multistage compressor many cylinders used in parallel for compression and after that the compressed air or gas is sent to the storage tank or to the utility end. Sometimes, discharge at high pressure is required and it is practically not possible to provide discharge at that pressure, reason being the excessive discharge pressure or other limitations. For such requirements two stage or more than two stage compressor is used. Inter-stage cooler may be provided to cool the air/gas before sending it to another stage for compression.

7.1.2 Rotary Compressor

Rotary compressor consists of rotor and the rotor rotates at high speed. They are compact in size. Screw compressor is an example of rotary compressor. It consists of a pair of helical screw which rotates in opposite direction and trap air to compress it. Rotary compressors are of continuous duty type and air or water cooled. Rotary air compressor is more popular for medium range of operation for its simple design and easy maintainability.

7.1.3 Centrifugal Compressor

Centrifugal compressor is dynamic compressor. As mentioned earlier also that the momentum is generated in the air due to rotation of rotor and that momentum is converted into pressure by diffuser. The centrifugal air compressor is suitable for large application especially where oil free air is required. It may have water cooled inter-stages.

7.2 Compressor Assessment

One of the parameter which is required to access the capacity of the compressor is measurement of FAD (Free Air Delivery). Free air delivery is the volume of flow of air/gas compressed and delivered at conditions of temperature, pressure and composition prevailing at compressor inlet. Deviation in FAD due to aging, poor maintenance etc. results in running of compressor for more time resulting in more energy consumption.

7.2.1 Capacity Assessment

The compressor under test is isolated from the system along with its receiver by closing the isolation valve. Initial pressure of the receiver is noted. The compressor is started and the stop watch is activated. The time taken to attain the normal pressure is noted. The air temperature is also noted.

The free air delivery is calculated from the following formula:

$$Q = (P_2 - P_1/P_0) \times V/T \text{ Normal m}^3\text{/Minute}$$

Where

P_2 = Final pressure after filling (Kg/cm^2)
P_1 = Initial pressure after filling (Kg/cm^2)
P_0 = Atmospheric pressure (Kg/cm^2)
V = Storage volume in m^3 including volume of delivery piping system
T = Time to take build up pressure to P_2

If the discharged air temperature is higher than the ambient then the above formula is multiplied by the following factor:

$(273 + t_1)/(273 + t_2)$

Where t_1 is the ambient air temperature and t_2 is the discharged air temperature.

7.3 Compressor Efficiency

7.3.1 Isothermal Efficiency

Isothermal efficiency can be calculated by the following formula:

Isothermal power = $P_1 \times Q_1 \times \log_e R/36.7$

P_1 = Absolute intake pressure Kg/m^2
Q_1 = Free air delivery in m^3/hr
R = Pressure ratio P_2/P_1

7.3.2 Volumetric Efficiency

Volumetric efficiency can be calculated from the below mentioned formula.

Volumetric efficiency = Free air delivery/Compressor displacement

$$= \pi/4\,(d^2 \times l \times s \times \tau \times n)$$

d = Cylinder bore, metre
l = Stroke, metre
s = Speed in rpm
τ = 1 for single acting and 2 for double acting
n = Number of cylinders

7.3.3 Specific Power Consumption

The most practical way to find out the compressor efficiency is to calculate the specific power consumption.

Specific power consumption = KW/Volume flow rate

We can compare the specific power consumption for different compressor provided by different vendors.

7.4 Leak Test

The compressed air system consists of receiver tank, piping, isolation valves, fluid traps, intermediate storage tanks, heaters etc. Leak can take place from any part of system. Most common places from where leak can take place are pipe joints, couplings, hoses, tubes, flanges, and valves.

Leakages increase with the system pressure, therefore to quantify and arrest the leakage is very important.

7.4.1 Quantification of Leakage

Leakage can be quantified by the following method.

The compressed air operated equipment is stopped and the system is charged to operating pressure by running the compressor. The loading and unloading time is noted. This exercise is repeated over eight to ten cycles for accuracy in measurement. The following formula is used to calculate the leak.

Leakage in percentage = $(T \times 100)/(T + t)$

$S = (T \times Q)/(T + t)$

S = System leakage in m^3/minutes
T = Time for which compressor runs on load (minutes)
t = Time for which compressor runs on no load (minutes)
Q = Compressor capacity in m^3/minutes

7.5 Steps of Energy Conservation in Compressed Air System

- The intake air of compressor has a significant effect on the input energy of a compressor.

- The intake air should not be warm since hot air needs excess energy to compress and according to a rough estimate every 4°C rise in temperature results in 1% increase in energy consumption. Therefore, necessary arrangement should be made to cool the intake air.
- The dirty air filters also result in excess energy consumption. Therefore, cleaning of air filters should be part of maintenance schedule.
- For compressors of larger size where compression takes place in multistages. Inter-stage cooler and after stage cooler are provided to cool the air or gas after every stage. Condensation takes place in inter-stage cooler. Therefore, condensate separator is provided before air re-enters in another stage. Coolers and condensate separators should work properly and there should not be any leakages.
- One of the important aspects of energy conservation in compressor is pressure setting. The performance of the compressor is optimum at a certain pressure and operation at higher as well as low pressure always results in poor efficiency and more energy consumption. The high and low pressure requirement should be segregated.
- The system should be designed in such a way that the pressure drop should be minimum. Care should be taken at design stage that the size of pipe lines, couplings, length of pipes etc. should be optimum.
- If compressors of many ratings and sizes are connected to a common header, compressors should be operated in such a way that cost of compressed air is minimum. In such a situation the compressor of lower rating should take care of the fluctuation in load and other compressors should work at their optimum capacity. On the other hand if all the compressors are of same rating then any one of the compressor should take care of the varying load.
- The important factor is arrest of leakages as mentioned earlier in this chapter.

7.6 Conclusion

As mentioned earlier also that compressed air system is such an area which required to be analysed critically from energy conservation point of view. Based on the analysis, various methods as mentioned in this chapter shall be adopted to minimize the cost of compressed air.

7.7 Software

The software has provisions for analysis of compressed air system.

7.7.1 Objective

The objectives of the software are:

- To prepare a database of the compressors for which performance analysis is required.
- Retrieval of data from the database based on certain criteria i.e. query generation.
- Transferring the retrieved data to the analysis sheet.
- Analysis of various parameters related with energy saving on analysis sheet.
- Saving and printing the results.

7.7.2 Procedure

To perform the above mentioned tasks, step by step procedure is mentioned in the following paragraphs.

- The first task is to prepare a database of the compressors for which energy auditing is to be done. To accomplish this, click on "Electrical Equipments: Inventory" of the main menu and then on "Compressor" of the submenu as shown in Fig. 7.0 by an arrow.

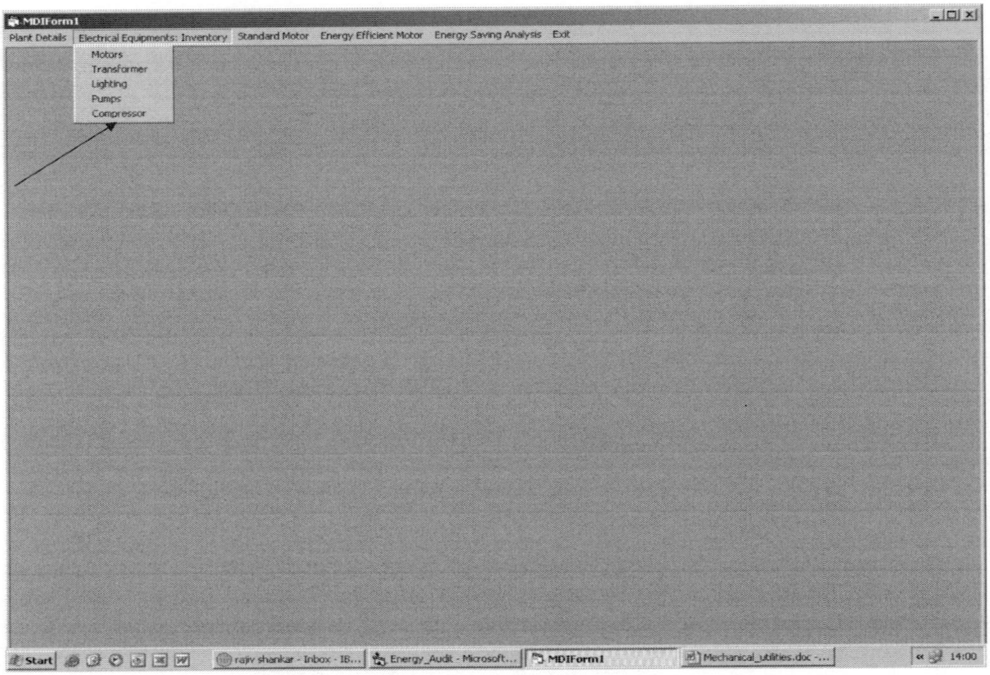

Fig. 7.0 Menu for compressor

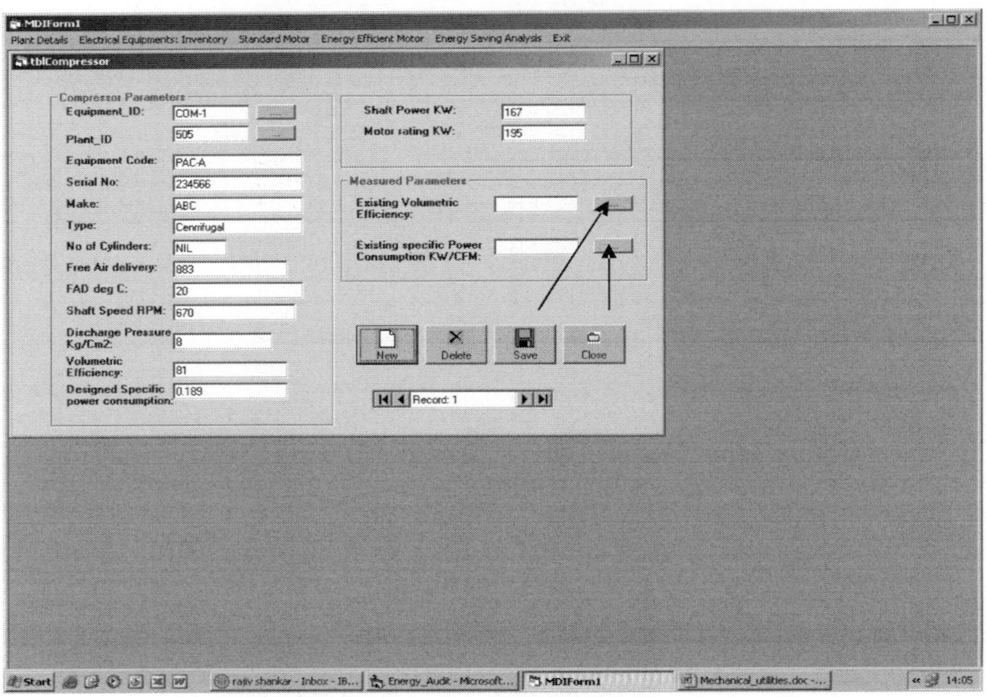

Fig. 7.1 Screen for data entry for compressor

A screen as shown in Fig. 7.1 will appear. Fill in the required parameters as it was done in earlier chapters while preparing database for other equipments. Some of the parameters can be obtained from the datasheet of the manufacturer. Existing volumetric efficiency and existing specific power consumption are to be calculated. Click on the button provided besides the box titled as "Existing Volumetric Efficiency" (as shown by an arrows in Fig. 7.1). This will result in opening up of a screen as shown in Fig. 7.3 where volumetric efficiency can be calculated. Similarly, specific power consumption can also be calculated.

- Click on "Energy Saving Analysis" of the main menu and then on "Compressor Analysis" of submenu as shown in Fig. 7.2 by an arrow.

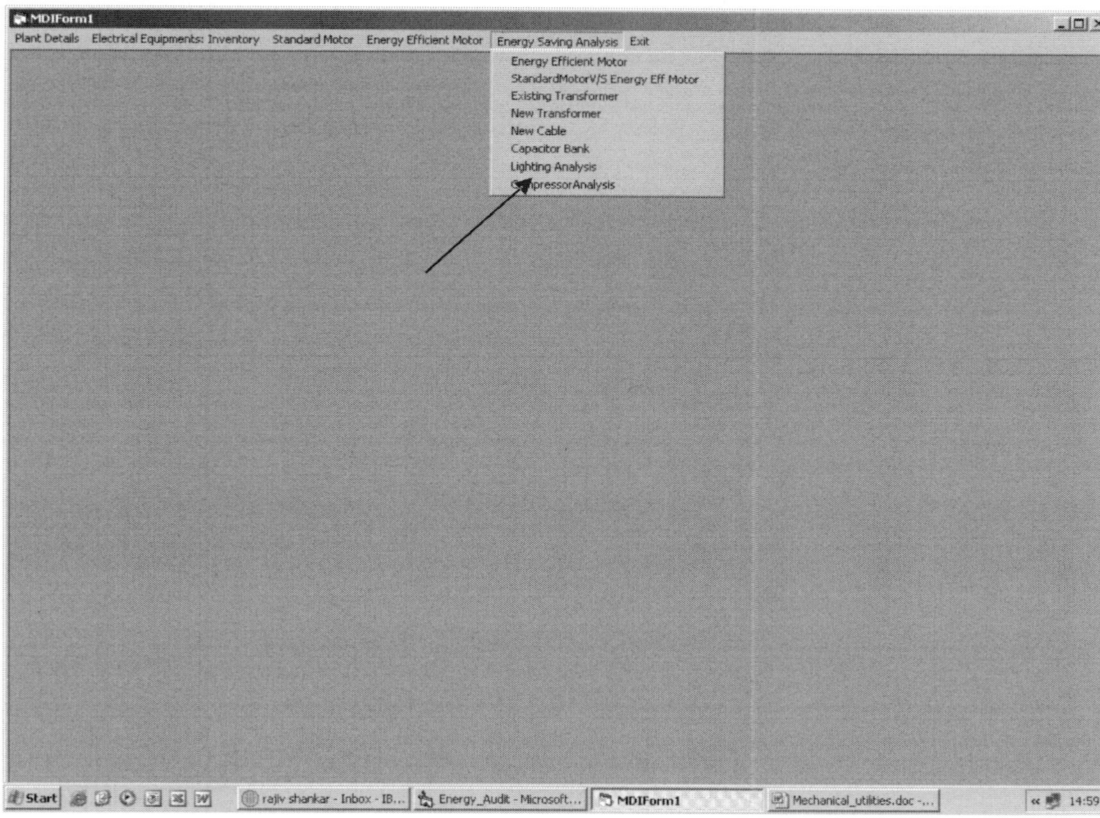

Fig. 7.2 Menu for analysis of compressor

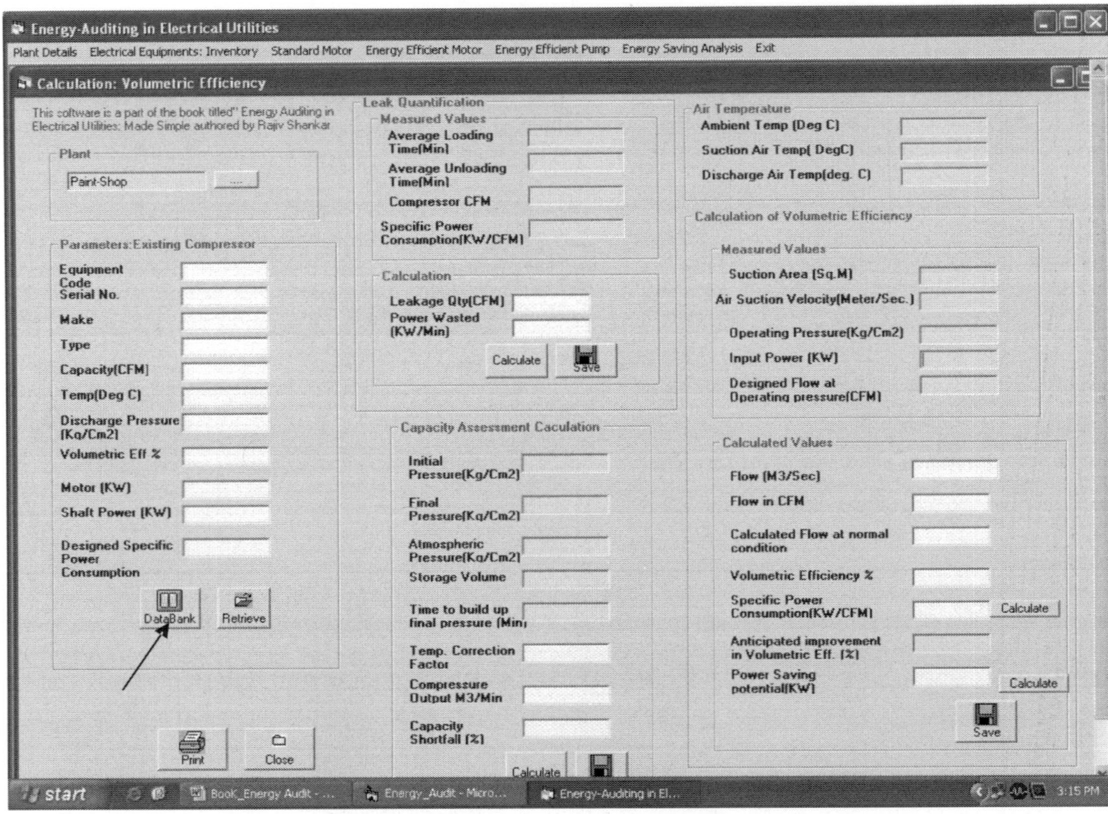

Fig. 7.3 Analysis for energy saving

A screen as shown in Fig. 7.3 opens up. Click on "Data Bank" button (as shown by an arrow in Fig. 7.3), it will open a screen as shown in Fig. 7.4. This is a search screen and plant name capacity and the power rating are to be specified as search criteria for retrieving the data from the database. Click on the "Search" button to retrieve the records. As shown in Fig. 7.4, two records are selected from the data bank in the search result table. To select a compressor from the table, click on the "Compressor-ID" as shown by an arrow in Fig. 7.4. By clicking on "COM-2" second record will be selected and all the data will automatically transferred on the analysis screen . Click on "Retrieve" button as shown by an arrow in Fig. 7.5 to retrieve all the detail on the screen.

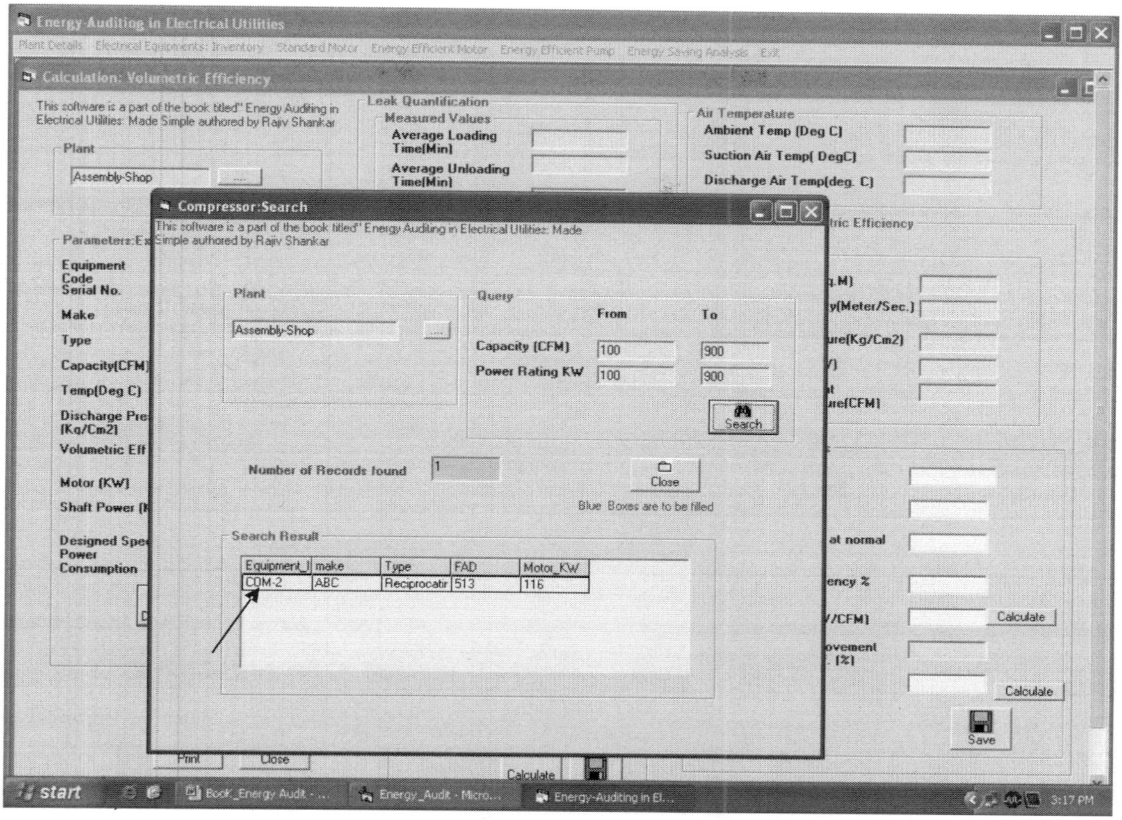

Fig. 7.4 Search by query generation

Various calculations like leak test; calculation for volumetric efficiency, calculation for capacity assessment of the compressor can be performed on this screen. The results of calculations can be printed by clicking on "Print" button as shown in Fig. 7.9.

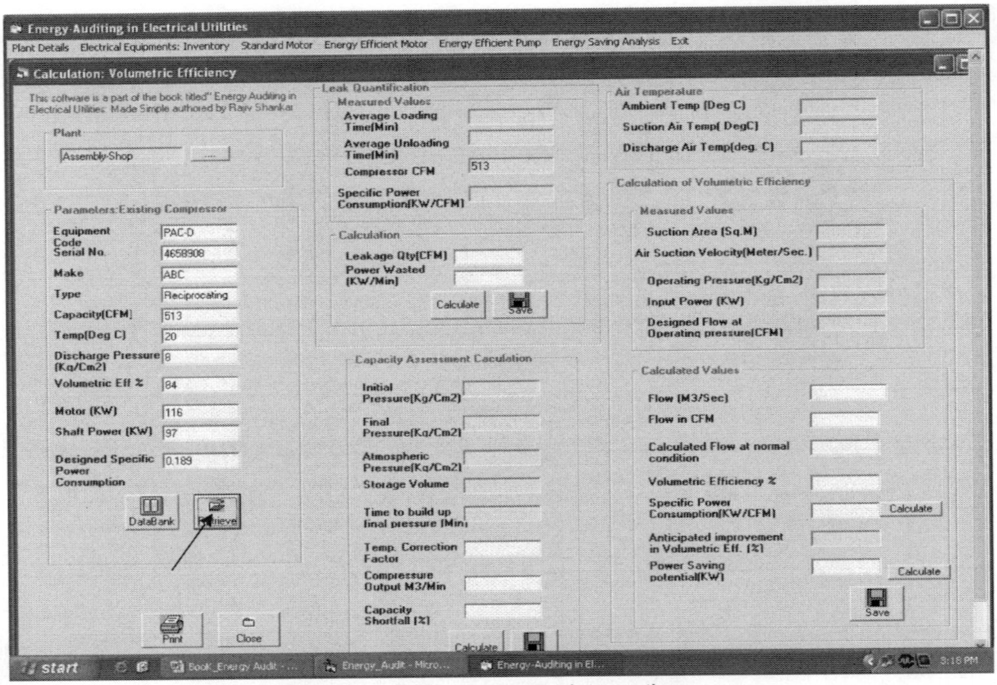

Fig. 7.5 Retrieval of compressor data on the screen

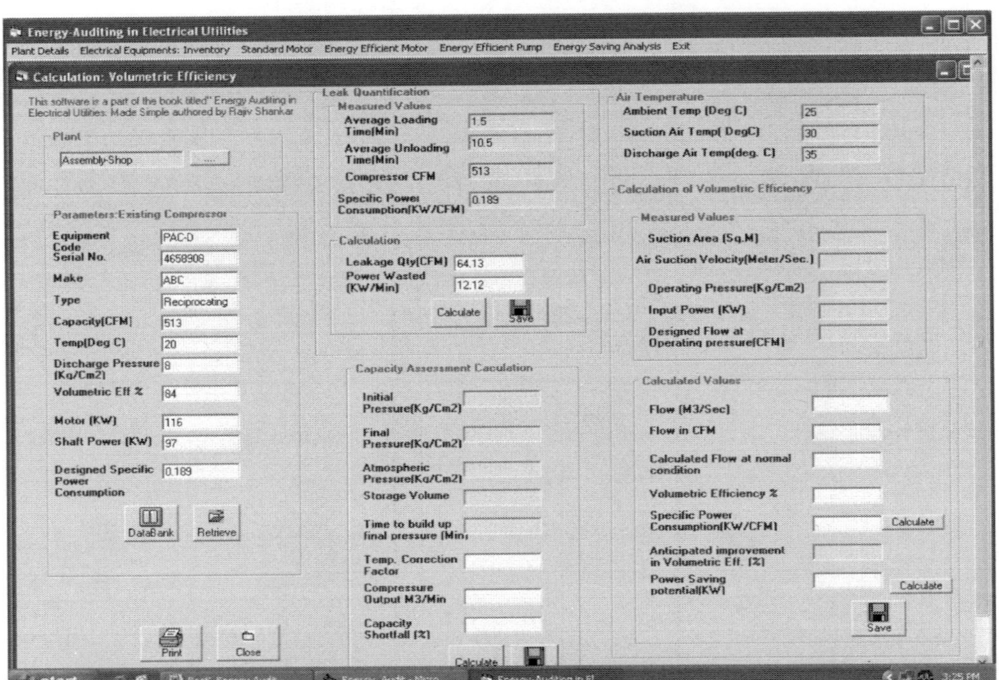

Fig. 7.6 Leakage calculations

Let's take an example.

Example 7.1

The following are the measured values.

Average loading time = 1.5 minutes

Average unloading time = 10.5 minutes

Compressor CFM = 513

Specific power consumption = 0.189 KW/CFM

Leakage quantity = 513 × 1.5/(1.5 + 10.5)

= 64.13 CFM

Power wasted per minute = 64.13 × 0.189

= 12.12KW

Example 7.2

Calculation for compressor capacity.

Initial pressure = 0.5 Kg/cm^2

Final pressure = 7.0 Kg/cm^2

Atmospheric pressure = 1.026 Kg/cm^2

Storage volume = 8.2 m^3

Time to built up pressure = 4 minutes

Ambient temperature = 25°C

Capacity of compressor = 513 CFM

Temperature correction factor

= (273 + 25)/(273 + 35)

= 0.97

Compressor output

= (7.0 – 0.5) × 8.2/(1.026 × 4)

= 12.99 m^3/minutes

= 12.99 × 35.32

= 458.71 CFM

Temperature correction

= 0.97 × 458.71 = 444.95 CFM

Capacity shortfall

= 100 × (513 – 444.95)/513

= 13.27%

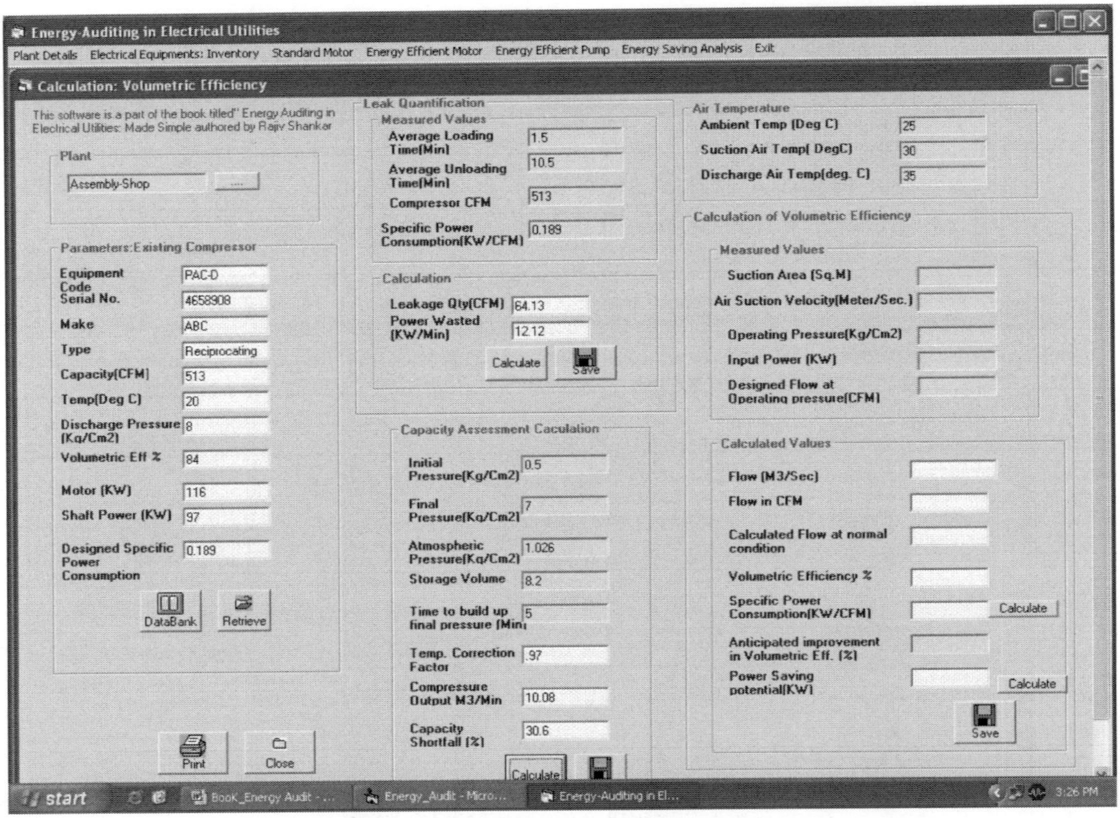

Fig. 7.7 Capacity shortfall calculation

Example 7.3

Volumetric efficiency calculations.

Suction area = 0.0072 m^2

Average air suction velocity = 30.5 m/sec

Suction air temperature = 30°C

Rated capacity = 513 CFM

Flow at operating condition

$= 30.5 \times 0.0072$

$= 0.22$ m^3/sec.

$= 0.22 \times 2118.87$

= 466 CFM

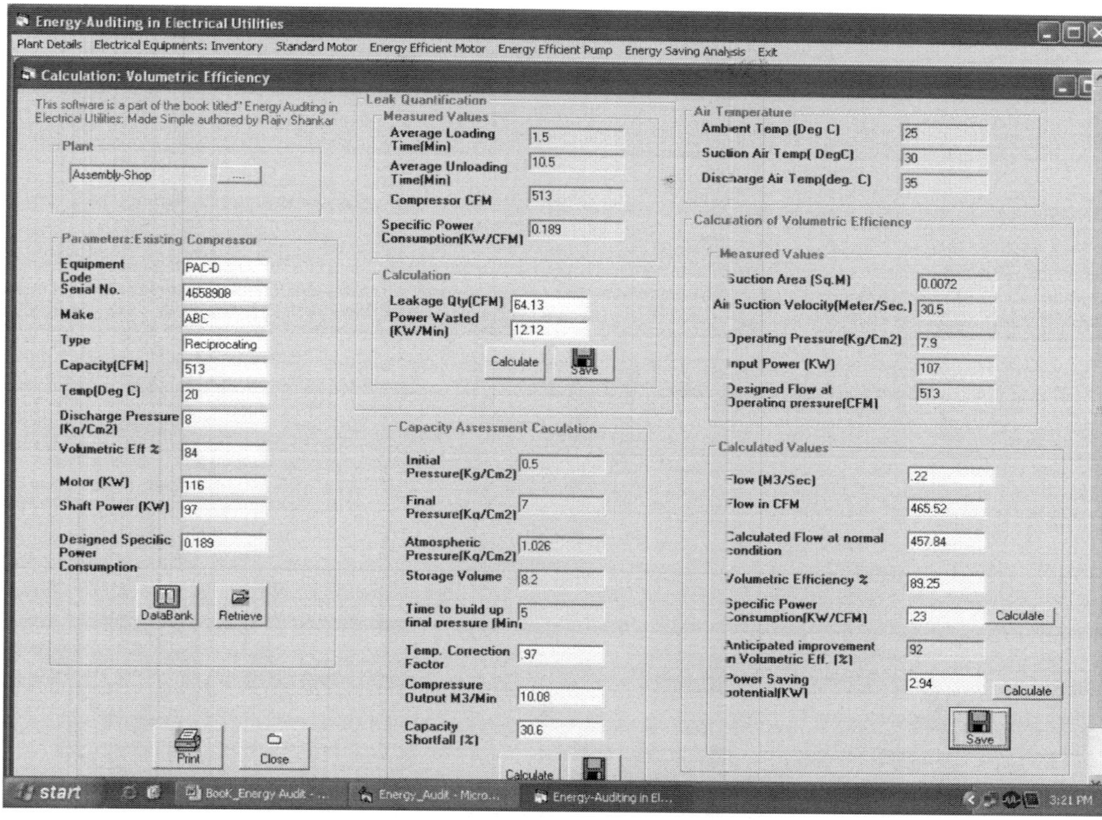

Fig. 7.8 Calculation of volumetric efficiency

Temperature correction factor

$= (273 + 25)/(273 + 30)$

$= 0.98$

Calculated flow at normal condition

$= 0.98 \times 466$

$= 456$ CFM

Volumetric efficiency

$= 456/513$

$= 88.89\%$

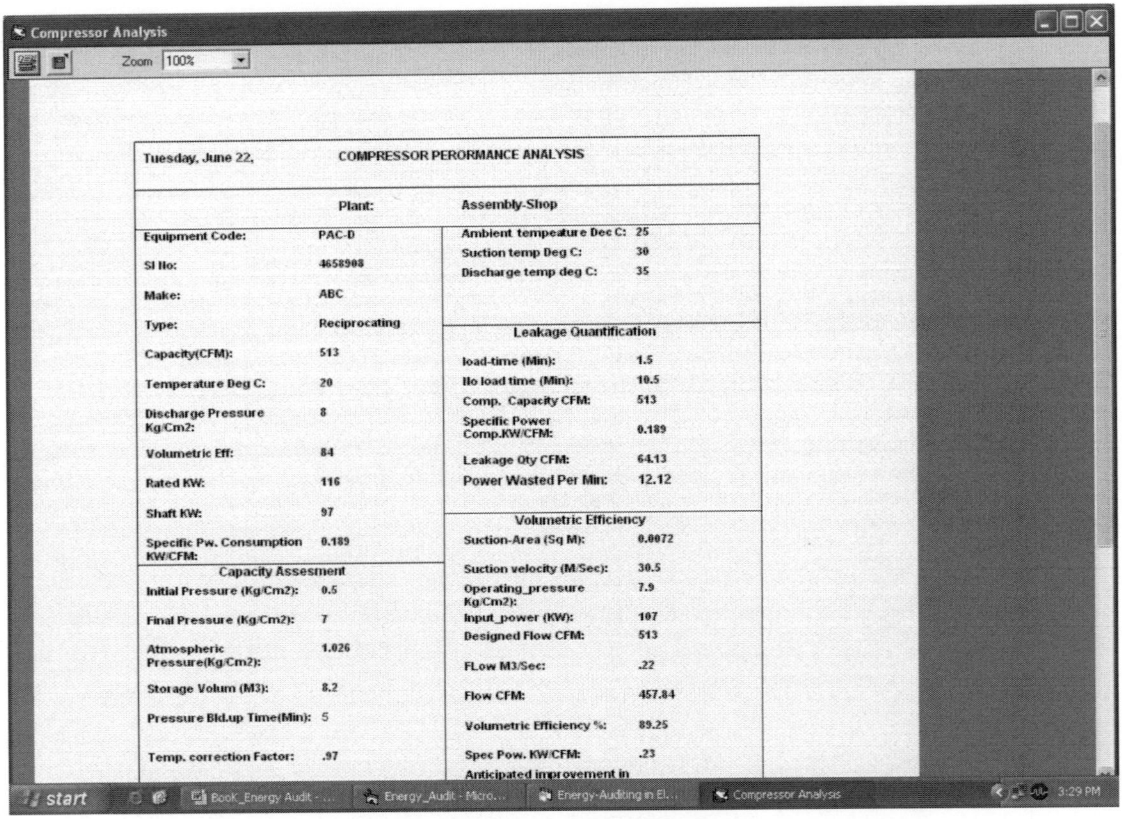

Compressor Analysis

Zoom 100%

Tuesday, June 22, COMPRESSOR PERORMANCE ANALYSIS

Plant: Assembly-Shop

Equipment Code:	PAC-D	Ambient tempeature Dec C:	25
Sl No:	4658908	Suction temp Deg C:	30
Make:	ABC	Discharge temp deg C:	35
Type:	Reciprocating	**Leakage Quantification**	
Capacity(CFM):	513	load-time (Min):	1.5
Temperature Deg C:	20	No load time (Min):	10.5
Discharge Pressure Kg/Cm2:	8	Comp. Capacity CFM:	513
Volumetric Eff:	84	Specific Power Comp.KW/CFM:	0.189
Rated KW:	116	Leakage Qty CFM:	64.13
Shaft KW:	97	Power Wasted Per Min:	12.12
Specific Pw. Consumption KW/CFM:	0.189	**Volumetric Efficiency**	
Capacity Assesment		Suction-Area (Sq M):	0.0072
Initial Pressure (Kg/Cm2):	0.5	Suction velocity (M/Sec):	30.5
Final Pressure (Kg/Cm2):	7	Operating_pressure Kg/Cm2):	7.9
Atmospheric Pressure(Kg/Cm2):	1.026	Input_power (KW):	107
Storage Volum (M3):	8.2	Designed Flow CFM:	513
Pressure Bld.up Time(Min):	5	FLow M3/Sec:	.22
Temp. correction Factor:	.97	Flow CFM:	457.84
		Volumetric Efficiency %:	89.25
		Spec Pow. KW/CFM:	.23
		Anticipated improvement in	

Fig. 7.9 Printing of calculation result

CHAPTER

8

Pumps

8.0 Introduction

A wide variety of pumps of different sizes for different application can be seen in a process industry. The most commonly used pumps are centrifugal and reciprocating type.

Depending on the working principle, pumps can be divided into two types. One is Dynamic or Displacement Pump and another is Rotary Pump. The reciprocating or positive displacement pumps are more efficient in comparison of centrifugal pumps. However, reciprocating pump needs more maintenance.

8.1 Centrifugal Pumps

The centrifugal pump mainly consists of two parts i.e. Impeller and Diffuser. Impeller is a rotating part and attached with the pump shaft. Pump shaft is coupled with the rotor of electrical motor. The impeller is housed in the diffuser. The impeller is made of various materials like cast iron, poly carbonate, stainless steel etc.

Water or any fluid for which pumping is required, enters from the centre of the impeller, sometimes it is called Eye and leaves the impeller due to the centrifugal force produced by the impeller during rotation. Due to this, a low pressure area is created near the eye of the impeller resulting in more liquid to flow into the pump. As liquid flows through the impeller, it develops velocity. This velocity is converted into pressure by the diffuser and specially designed passageway of the pump. This passageway directs the liquid towards the discharge of the pump.

The pressure or the pump head depends on the following:

- Impeller diameter
- Numbers of impeller
- Size of impeller eye
- Impeller speed

Capacity of pump depends on the head and flow rate of the liquid. If the head and flow are more then the capacity of the pump will be more and more input power is required. According to the size of pump capacity, motor is selected. The amount of water/liquid to be pumped by the centrifugal pump depends on the pressure against which liquid is to be pumped. If pressure of head is more then flow will be less. Since pump is a dynamic device, the pressure can be considered in term of head i.e. meters of water column.

8.2 Characteristic Curves of a Centrifugal Pump

Fig. 8.0 shows the characteristic curves of a centrifugal pump.

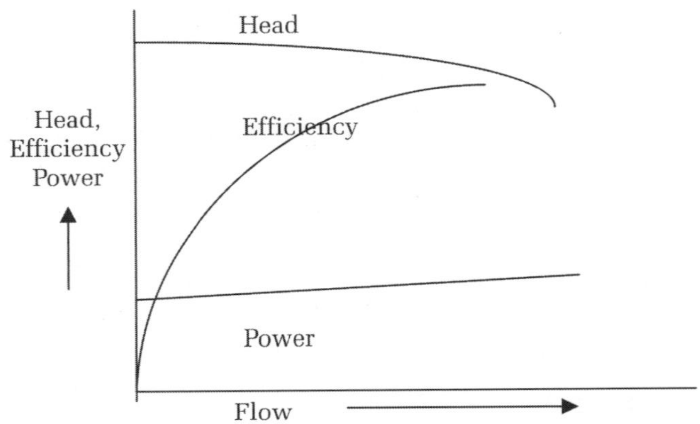

Fig. 8.0 Pump characteristic

8.3 Pump Performance Calculations

8.3.1 Pump Hydraulic Power = $Q \times H_{d-} H_s \times \rho \times g/1000$

Q = Flow, m^3/sec
H_d = Discharge head metre
H_s = Suction head metre
ρ = Fluid density
g = Acceleration due to gravity m/sec^2

8.3.2 Pump Shaft Power

= Hydraulic power/Pump efficiency

8.2.3 Electrical Input Power

= Pump shaft power/Motor efficiency

8.4 System Curve

A liquid has to overcome some resistance of that system to flow in that system. The resistance results in loss of head. The loss of head is due to the following:

- Difference between the height of supply and storage reservoirs. The liquid is pumped against the gravity. The static head is not dependent on flow and graphically represented by a straight line.

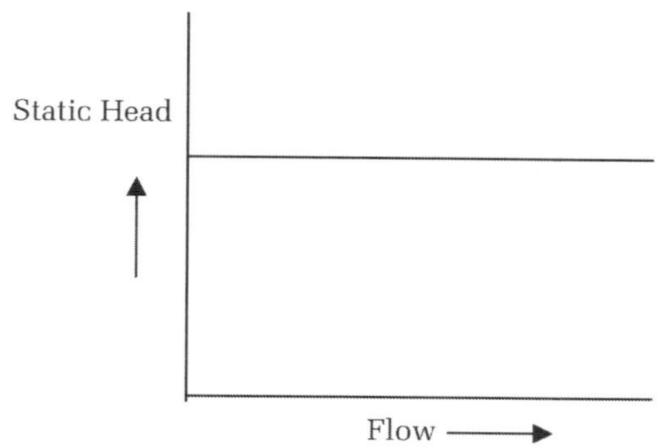

Fig. 8.1 Static head vs Flow characteristic

- Loss of head due to friction in the pipe lines, valves and equipment in the system. The friction head loss is proportional to square of the flow. The frictional head loss can be determined for different sizes to pipe lines and fittings with the help of friction loss tables. The system friction head loss vs flow curve is shown in Fig. 8.2.

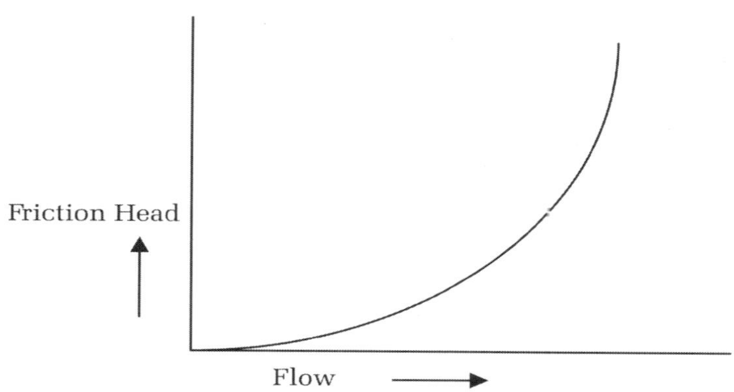

Fig. 8.2 Friction head vs Flow characteristic

Most of the system has both static and friction head loss and the system curve is the combination of these two curves. As shown in Fig. 8.3.

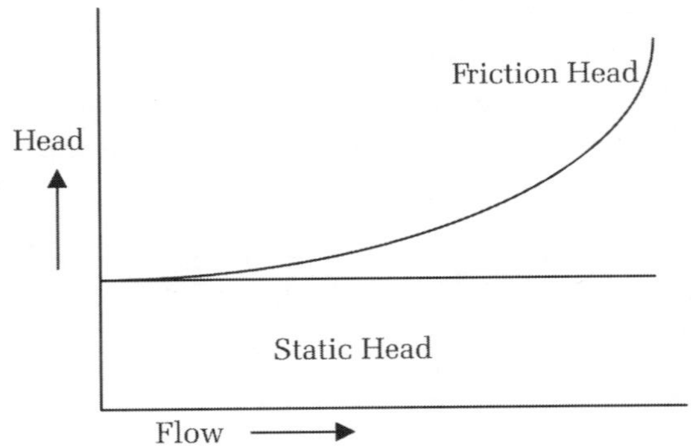

Fig. 8.3 System with static head

Static head depends on location where pump has been installed. Reduction in static head will result in reduction in pump sizing, so that both running and capital cost will get reduced. Similarly, friction loss depends on the sizes of the pipe lines, fittings and valves etc. Reduction in friction loss will reduce the running cost of pump.

8.5 Pump Characteristic

The head vs flow characteristic curve of a centrifugal pump is shown in Fig. 8.4. As seen from the curve, the head falls gradually as the flow increases. If the system curve is also drawn on the same graph, the two curve will intersect at one point and this point is called Operating Point of the Pump. The pump is selected so that the system curve and pump curve are matched in such a way that the pump operates at its maximum efficiency. Pump efficiency changes with change in flow rate and pressure or head. Due to physical change in the system like change in elevation, pipe diameter, pipe length, pipe fittings etc. flow changes and accordingly system curve also changes. Pump is selected in such a way that it matches with the system curve.

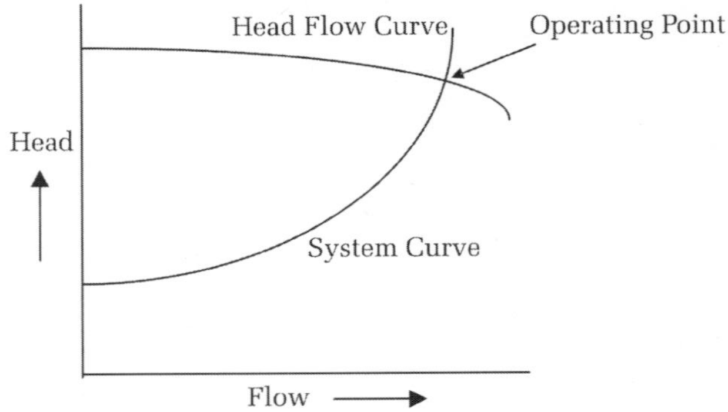

Fig. 8.4 Pump operating point

8.6 Sizing of Pump

Let's take an example. Water has to be lifted up to a height of H metre. Therefore, the static head will be H metre. System curve is shown in Fig. 8.5.

The flow rate is 300 m^3/hr and the pump operating point is A. Pump efficiency at this point is 80%. The head is 30 metres. Now, due to some reason the actual requirement of flow changes from 300 m^3/hr to 200 m^3/hr. This reduction in flow is achieved by throttling the valve in the line i.e. if the valve is opened more water will flow more and vice versa. Due to this act, the system resistance increases and the system curve shifts towards the upper side as shown in Fig. 8.5. The new operating point will be B. The head at the operating point B will be 50 metres and the flow will be 200 m^3/hr. The efficiency will get reduced to 70%.

Since the flow has reduced, the capacity of pump also needed to be reduced to meet the new requirement. The pump capacity can be changed by various means like, replacement of existing pump with a pump having smaller capacity, trimming of the impeller, reduction in pump speed by providing variable speed drive. This changes the characteristic of pump. As shown in Fig. 8.5, the operating point with new characteristic is C. It may be noted that at this point the head has reduced to 28 metre. The input power requirement will also get reduced by reduction of head resulting in saving in energy.

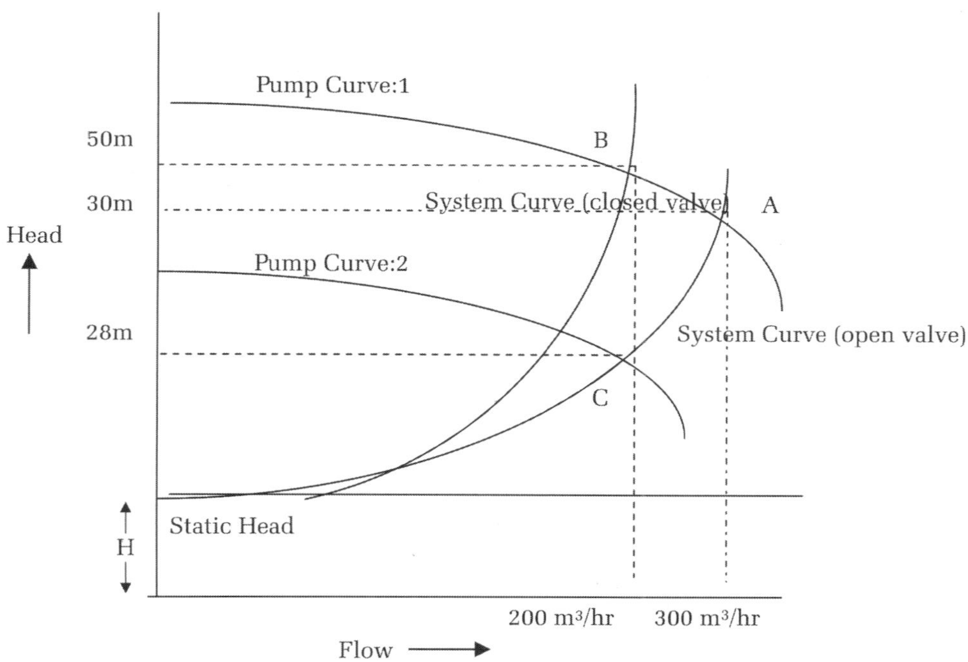

Fig. 8.5

8.7 Energy Saving in Pumping System

We may find that in many applications oversized pumps are provided and the flow is controlled by restricting the opening of the valve. The reason of providing oversized pumps may be high safety of margin during design stage. The flow requirement should be optimized and then accordingly the pump should be selected.

8.7.1 Speed Variation of Pump

The parameters like head, flow and power absorbed by the pump are related with the rotational speed of the pump impeller. The equations relating to flow, power and head with speed are known as Affinity Law.

As per affinity law

$Q \alpha N$

$H \alpha N^2$

$PW \alpha N^3$

Q = Flow

H = Head

PW = Power

N= Speed

Suppose a pump is operating at a speed of 1450 RPM (N_1)

Flow rate $Q_1 = 50\ m^3/hr$

Power absorbed $P_1 = 12KW$

Head = 50 metre

The speed increases to 2000 RPM (N_2)

The new flow rate will be:

$Q_1/Q_2 = N_1/N_2$

$50/Q_2 = 1450/2000$

$Q_2 = 50 \times 2000/1450$

$= 69\ m^3/hr$

The new head will be:

$H_1/H_2 = (N_1/N_2)^2$

$50/H_2 = (1450/2000)^2$

$H_2 = 95.12$ metre

The new power requirement will be:

$P_1/P_2 = (N_1/N_2)^3$

$P_1/P_2 = (1450/2000)^3$

$P_2 = 31.5KW$

A small increase in speed results in large increase in power. If the speed is doubled then the power requirement will increase eight times. Similarly, a small decrease in speed causes a large reduction in power requirement.

8.7.2 Change in Impeller Diameter of Pump

The same affinity law as mentioned before also holds goods in case of change in impeller diameter also.

$Q \alpha D$

$H \alpha D^2$

$PW \alpha D^3$

Where, D is the impeller diameter. The reduction in the diameter of the pump impeller should be restricted to 75% of the diameter. Beyond that limit the efficiency will be affected. Besides this, if diameter is reduced, it will be permanently change. Due to any reason if flow increases in future then the impeller needs to be replaced. While in case of variable speed drive, the speed can be increased or decreased as per flow requirement.

8.7.3 Parallel Pumping

Parallel pumping is adopted where static head is high. The additional pump is switched on and off to meet the additional flow requirement. System curve is not affected by putting the pump operation in parallel. The head will be the sum of head of all the pumps operating in parallel for a particular flow. The operating points of the pumps running in parallel should remain within the region as per the manufacturer's requirement.

8.8 Conclusion

In process industry quite a number of centrifugal pumps are installed for various processes. As mentioned in this chapter that a small change in speed results in large amount of power saving. Therefore, the flow requirement will be optimized first and according to that speed of the pump should be controlled. For controlling the speed, motor with variable speed drive is the most suitable option. The motor speed is varied by changing the frequency of the supply feeding the motor. Smart motor controller and drives are available in the market with improved technology.

8.9 Software

Let's have a look into capability of the software for performing the energy saving analysis in pumps.

8.9.1 Objective

- To build a database of existing pumps for which energy auditing is to be carried out.
- To build a database of other energy efficient pumps.
- To retrieve data from the database.
- Transfer these retrieved data on the pump analysis sheet.
- Energy saving analysis.
- To take a printout of the results.

8.9.2 Procedure

To accomplish above mentioned objectives following steps are to be followed.

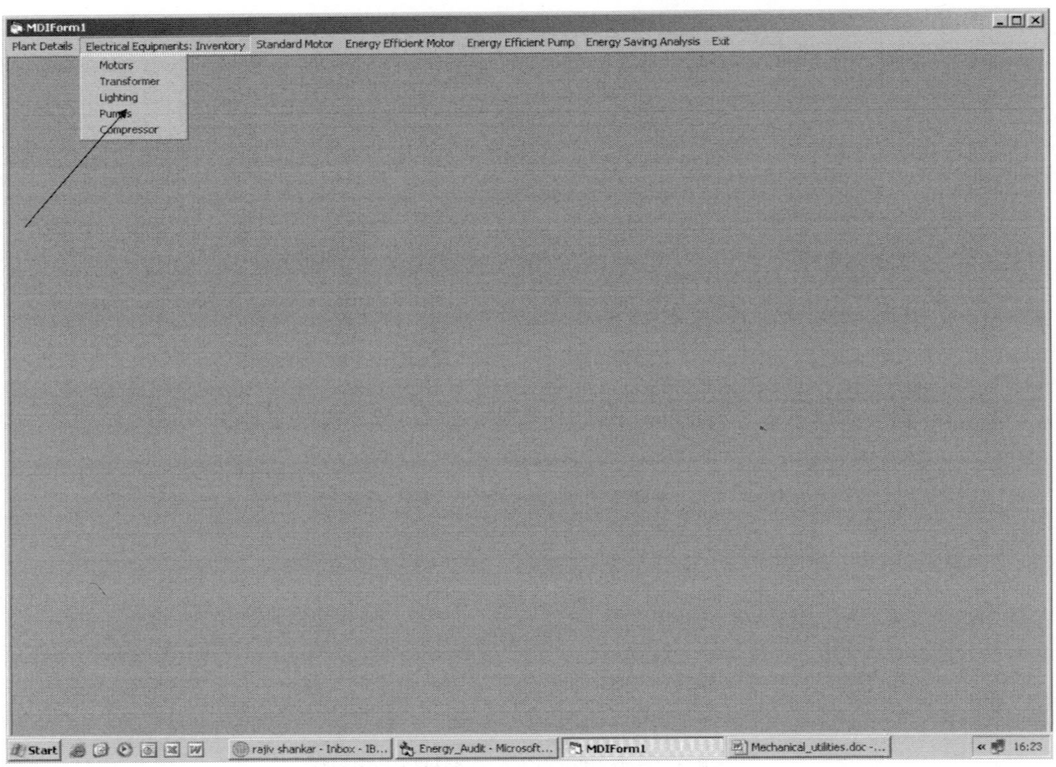

Fig. 8.6 Main screen

- Click on "Electrical Equipments: Inventory" and then click on "Pumps" as shown by an arrow in Fig. 8.6. A screen as shown in Fig. 8.7 will appear.

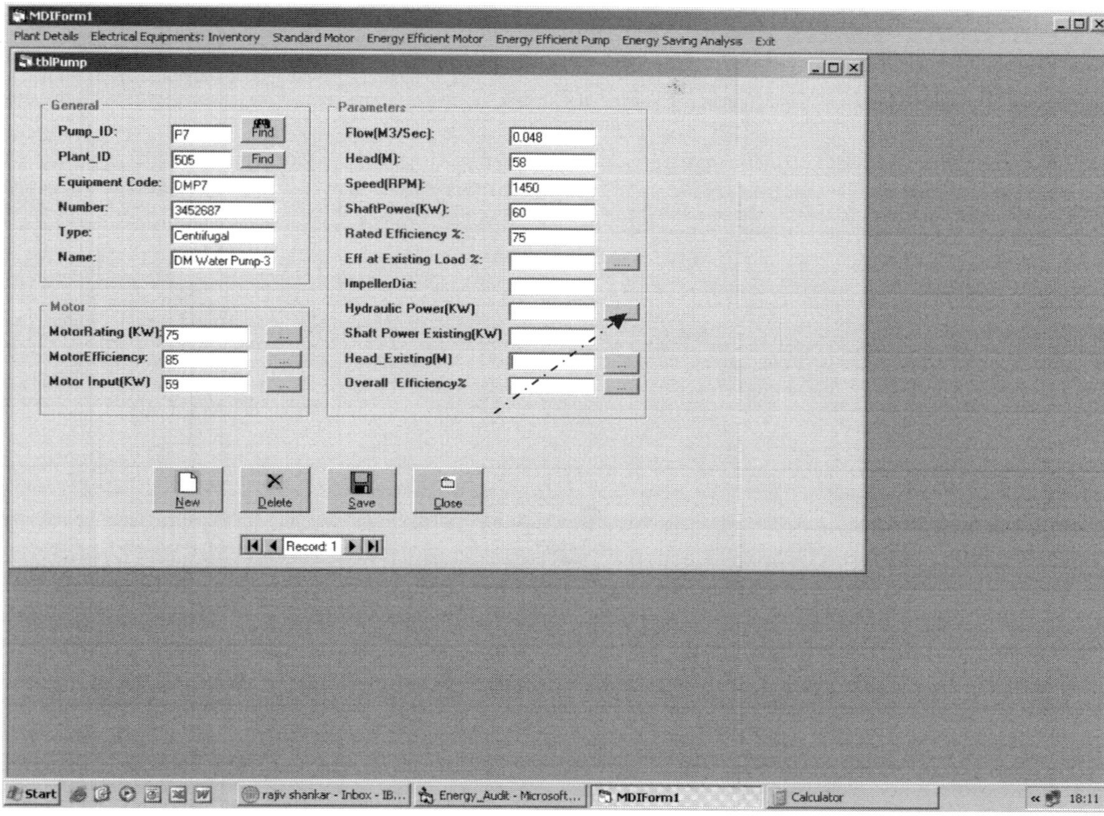

Fig. 8.7 Screen for storing existing pump information

This screen is for filling information related with the existing pump for making a database. Some of the information are to be obtained from the test certificates provided by the pump manufacturer, while other details can be obtained by doing measurements at site. Values from these measured parameters are calculated. For example, for obtaining pump efficiency, click on button as shown by dotted arrow in Fig. 8.7. On clicking this button will result in opening of another screen as shown in Fig. 8.8 where pump efficiency and hydraulic power can be calculated.

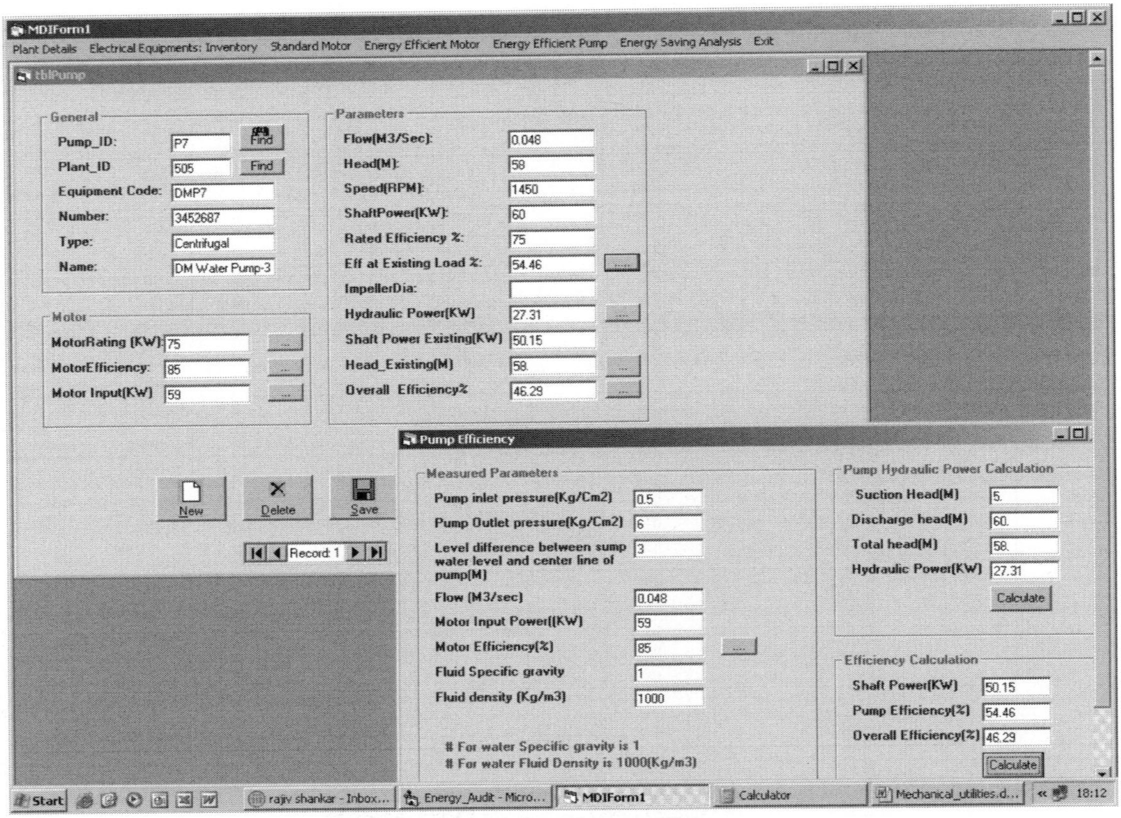

Fig. 8.8 Calculation of pump efficiency

Example 8.1

Calculation of hydraulic power, pump efficiency etc.

Measured parameters:

Pump inlet pressure = 0.5 kg/cm^2
Pump outlet pressure = 6.0 kg/cm^2
Level difference between sump level and centre line of pump = 3 metre
Flow = 0.048 m^3/sec.
Motor input power = 59KW
Motor efficiency = 85%
Fluid specific gravity = 1
Fluid density = 1000 Kg/m^3

Calculation:

$$\text{Head (H)} = \frac{(\text{Pump outlet pressure} - \text{Pump inlet pressure}) \times 10}{\text{Specific gravity of fluid}}$$

$$= \frac{(6.0 - 0.5) \times 10}{1} = \frac{(5.5) \times 10}{1} = 55 \text{ metre}$$

Total head = H + Level difference between sump level and centre line of pump

= 55+3 = 58 metres

$$\text{Hydraulic power} = \frac{\text{Head} \times \text{Flow } (m^3/\text{sec}) \times \text{Gravity } (9.8\ m^3/\text{sec}^2) \times \text{Fluid density}}{1000}$$

= 58 × 0.048 × 9.8 × 1000/1000

= 27.28KW

Shaft power = Motor input × Motor efficiency

= 0.85 × 59

= 50.15KW

Pump efficiency = 100 × Pump hydraulic power/Shaft power

= 27.28 × 100/50.15

= 59.39%

Overall efficiency = 100 × Pump hydraulic power/Motor input

= 27.28/59

= 46.23%

Same results are obtained by software calculation also as shown in Fig. 8.8.

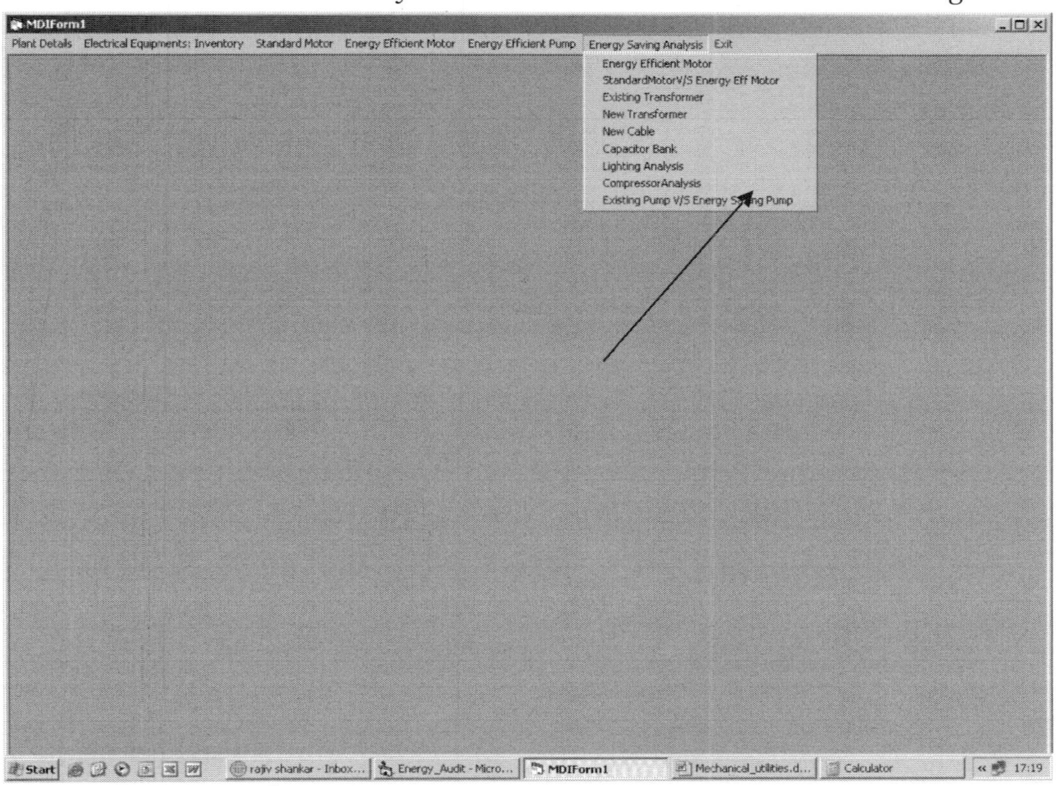

Fig. 8.9 Pump analysis menu

- Click on “Energy Saving Analysis” of main menu and also click on “Existing Pump vs Energy Saving Pump” of submenu. This action will open a screen as shown in Fig. 8.10.

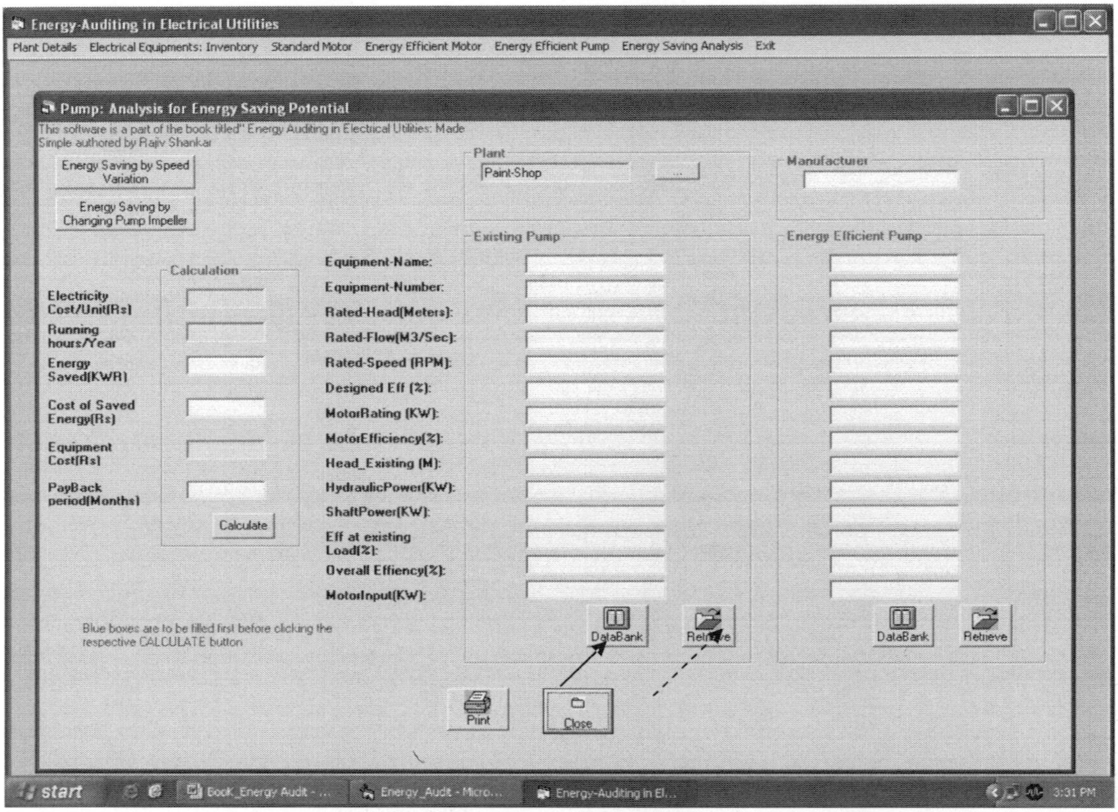

Fig. 8.10 Pump analysis sheet

Click on “Data Bank” button as shown by a solid arrow in Fig. 8.10. It will open another screen as shown in Fig. 8.11. This screen is for retrieving the details of the existing pump based on certain query like head, speed and efficiency. All the pumps which will fall under query range as mentioned in the query parameter boxes will be retrieved. For transferring these parameters on the main analysis screen, click on the “Pump-ID”. The query screen will disappear. Click on “Retrieve” button as shown by a dotted arrow in Fig. 8.10 all the parameters of the selected pump will be retrieved and copied in the respective boxes provided on the screen for the existing pump (Fig. 8.12).

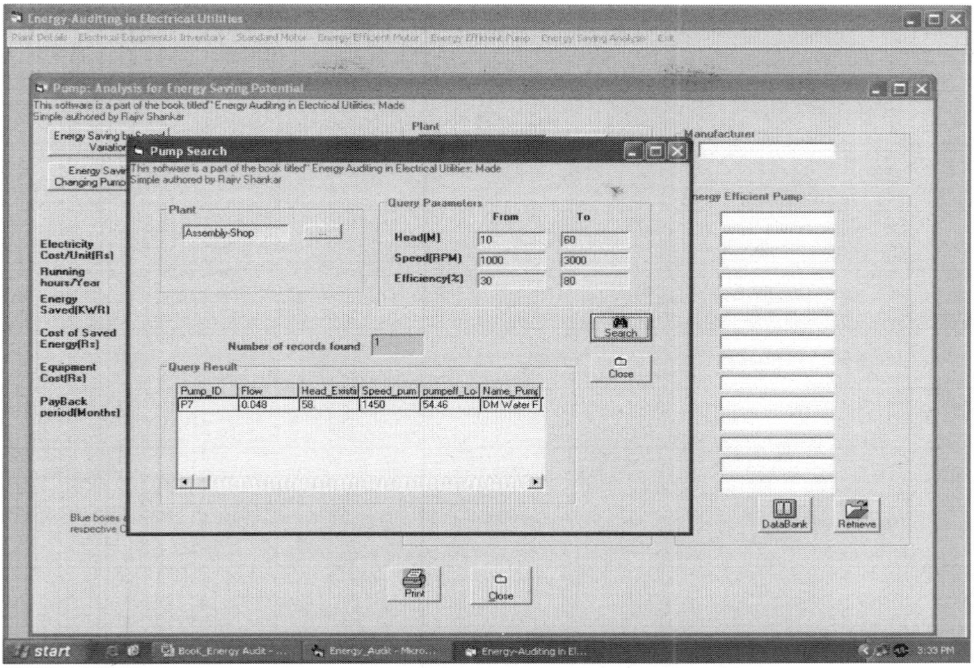

Fig. 8.11 Query generation for existing pump

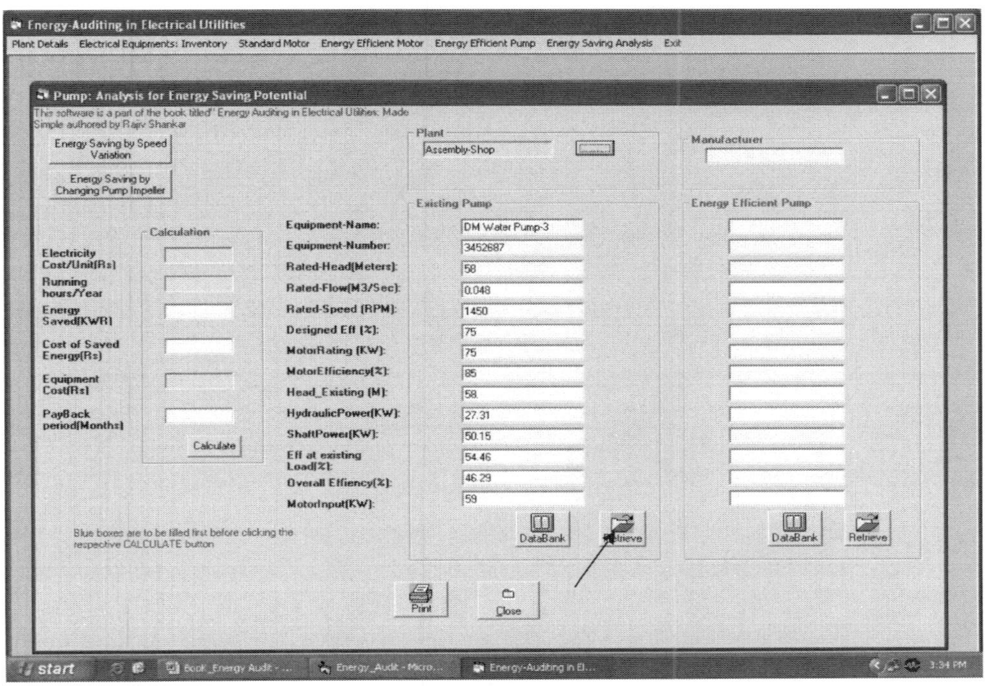

Fig. 8.12 Retrieval of existing pump parameters

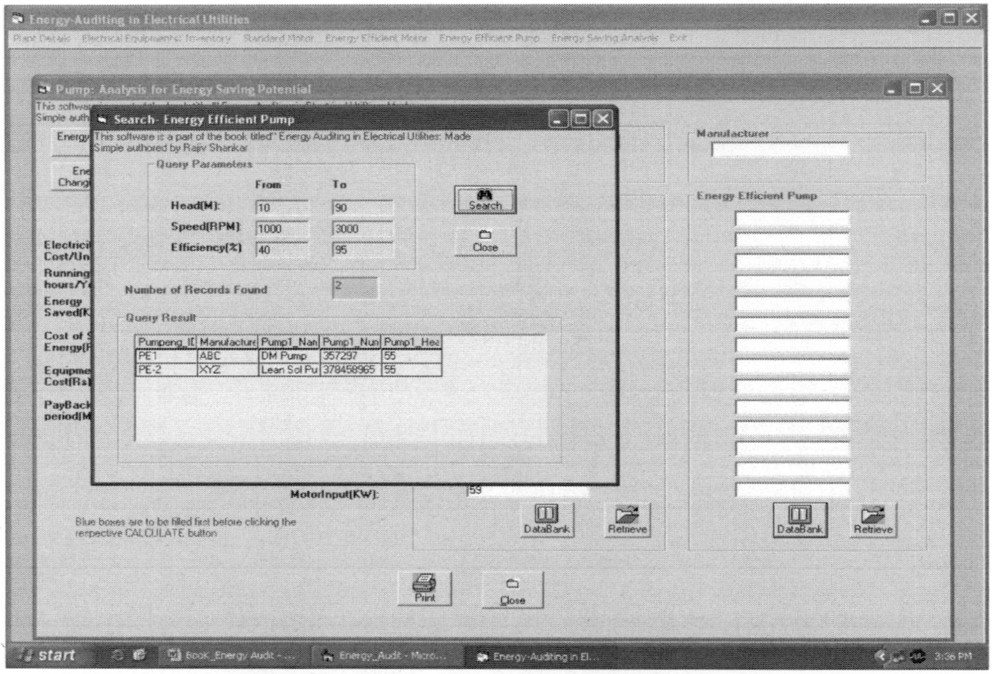

Fig. 8.13 Query generation for energy saving pump

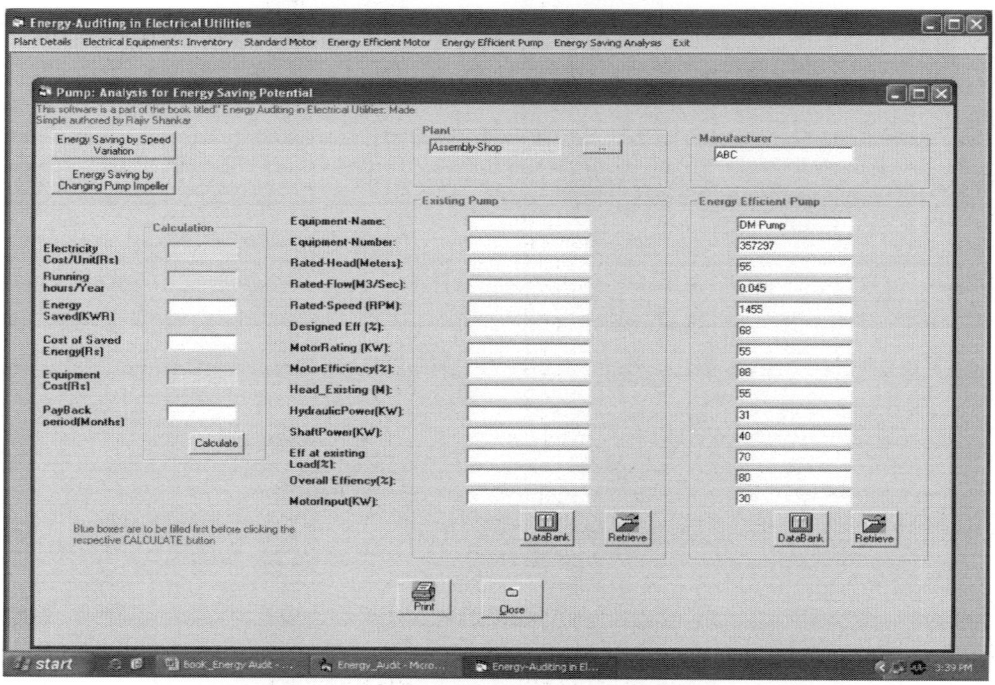

Fig. 8.14 Retrieval of data for energy saving pump

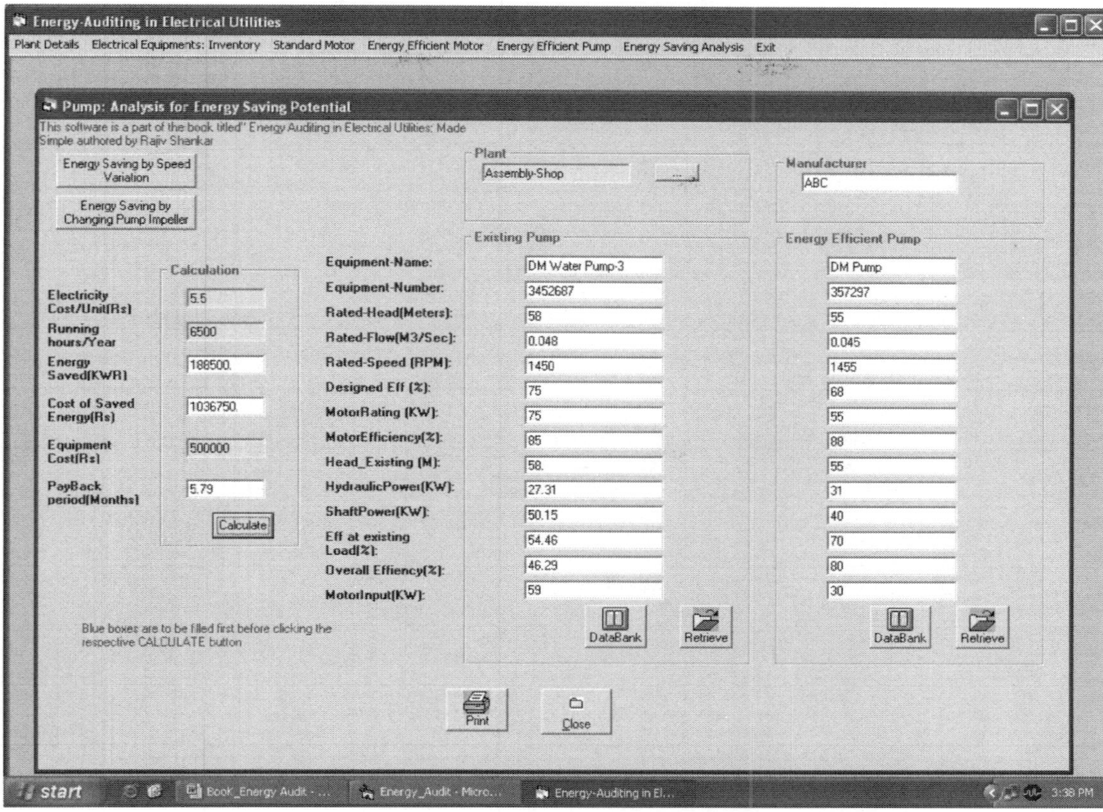

Fig. 8.15 Energy saving calculation

Similarly, data can be retrieved for the energy efficient pump also as shown in Fig. 8.13 to Fig. 8.15.

- Fill in the cost of energy, running hours for the pump and cost of new pump in the space provided in calculation frame as shown in Fig 8.15 and click on the "Calculate" button. Payback period, cost of saved energy will be calculated as shown in Fig. 8.16.
- Printout of calculation can be taken by clicking on "Print" button.

Example 8.2

Input power for the existing pump = 59KW
Input power for the new pump = 30KW
Running hours = 6500
Electricity cost = Rs5.5

Calculation for saving of energy cost.

Saved energy = 6500 × (59 – 30)
= 188500 KWH
Cost of saved energy
= 188500 × 5.5
= Rs10,36,750
Cost of new pump = 500000
Payback period
= 500000 × 12/1036750
= 5.78 months
App. 6 months

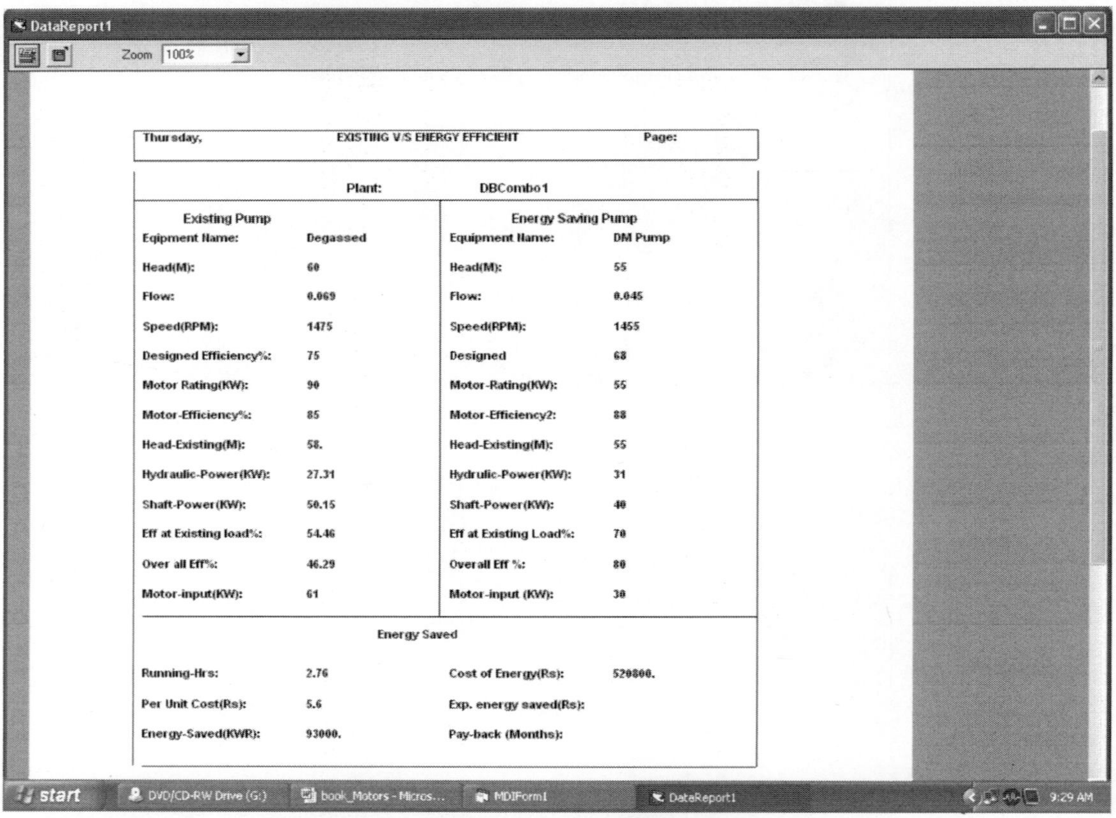

Fig. 8.16 Printout of calculated results

Calculation for energy saving by variation of speed and also by trimming of impeller can also be performed. Click on the buttons provided on the left hand side top corner of the screen as shown in Fig. 8.17 by a solid arrow. The screen as shown in Fig. 8.18 and

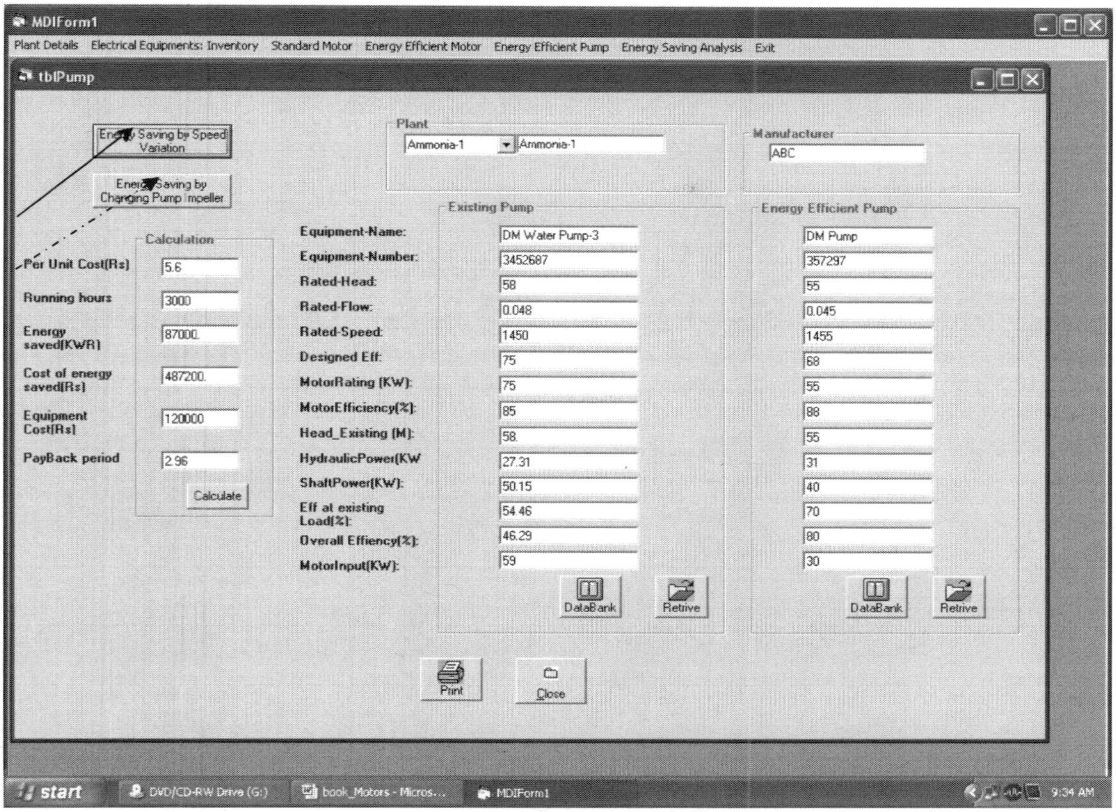

Fig. 8.17 Energy saving by speed variation

8.19 will open up by clicking on "Calculate" button. Before that the required values are to be filled in the respective boxes. The calculation results can also be printed by clicking on "Print" button as shown in Fig. 8.20.

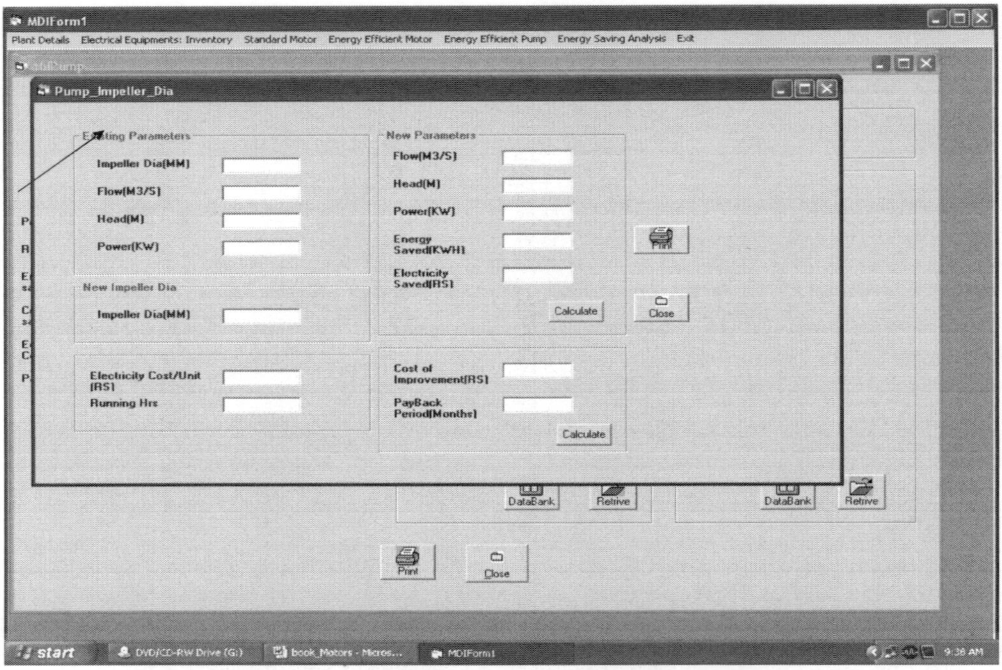

Fig. 8.18 Screen for calculation of energy saving by speed variation

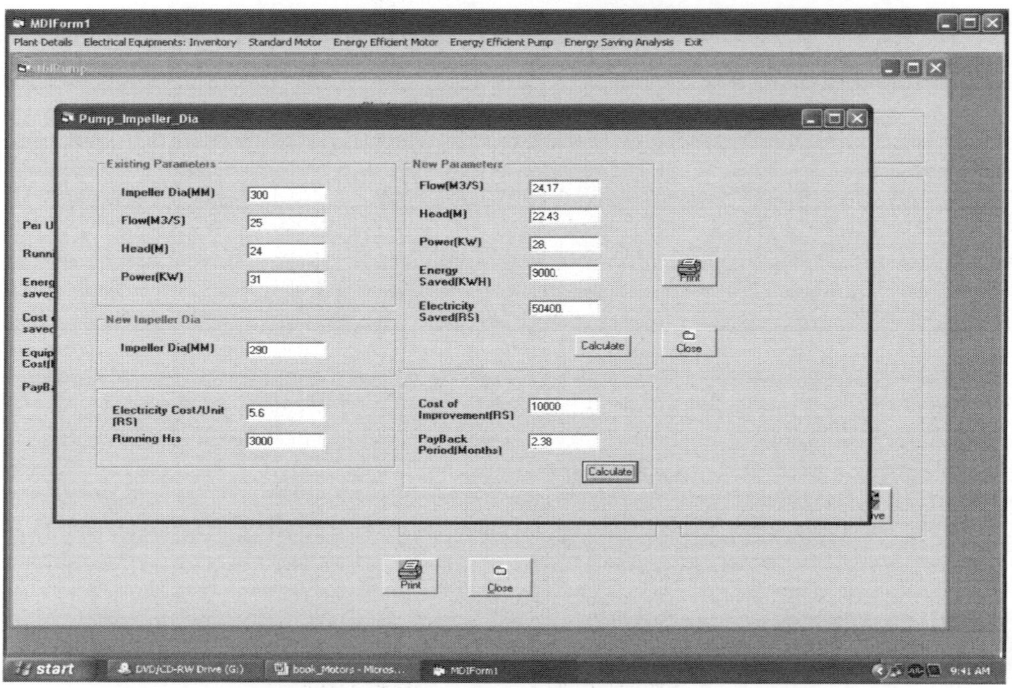

Fig. 8.19 Screen for calculation of energy saving by trimming of impeller

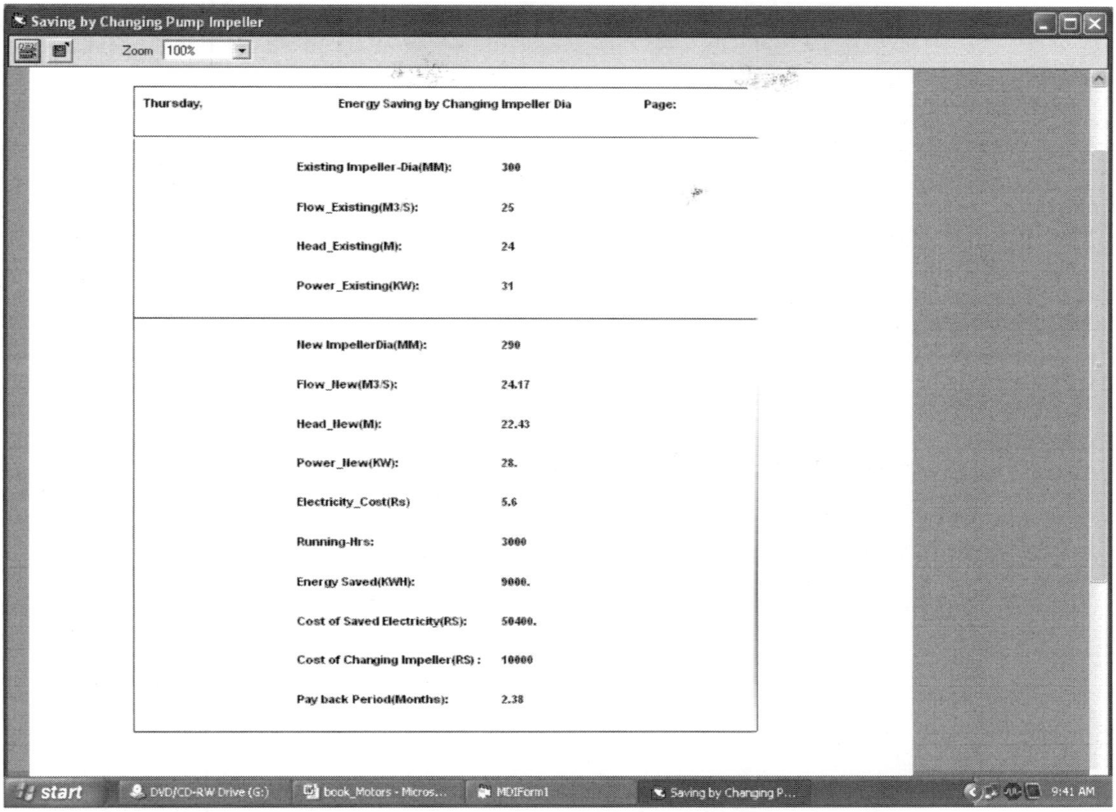

Fig. 8.20 Screen for printout of results for calculation of energy sav ng by trimming of impeller

CHAPTER

9

Air Conditioning System

9.0 Introduction

Air conditioners of various capacities ranging from a fraction of ton to several tons are installed in different type of establishment like domestic, commercial or industrial. According to the capacity of the air conditioning system, its construction also changes. For example, window type composite unit may work for small capacity of system, while for large system a package unit consisting of compressors, chillers, water cooling pipe lines, air handling units, cooling towers etc. are required. The air conditioning system is of two types.

- Vapour compression
- Vapour absorption

The most common type is vapour compression type of system.

Vapour compression system has following parts:

Air Circulation System

Air circulation system consists of air handling unit. Air handling unit has a coil in which cooled water circulates. Air blows over these coils and gets cooled. This cooled air goes back to the space/area which is required to be cooled through the inlet air ducting. Air blowers are used to circulate the air.

Chilled Water System

Chilled water that flows in the air handling units comes from the chiller. Water pumps are used to circulate this chilled water in the air handling units. The chilled water circulates through the air handling unit and the chiller.

Refrigerant System

In this system the state of refrigerant changes from vapour to liquid and liquid to vapour. Whenever refrigerant changes state from liquid to vapour, it takes heat from the surrounding and releases heat to the surrounding on changing state from vapour to liquid. The refrigerant is evaporated (converted into gaseous form) in chiller and takes heat of the water circulating in the chiller. This gaseous state is converted into liquid state by compressing the gas in compressor. When the gaseous state of the refrigerant is converted into liquid state the heat is released.

Condenser Water System

The liquid refrigerant is sent to condenser unit where water absorbs the heat. It's a type of heat exchanger. The cooled refrigerant which is still in liquid form is sent to chiller/expander unit by the pumps where it expands and converts into gaseous form and this cycle keeps ongoing.

Cooling Tower System

The hot condenser water is pumped to cooling tower where water flows down through the fills and air fan forces the air through the water. In this process water rejects the heat to the atmosphere.

9.1 Energy Efficiency in Air Conditioning System

The requirement for cooling is quantified as tons of refrigeration and can be calculated by the following formula:

TR = F × Sp × (Ti – To)/3024

F = Mass flow rate of the coolant in Kg/hr

Sp = Specific heat of the coolant in K cal/Kg °C

Ti = Inlet temperature of the coolant to chiller in °C

To = Outlet temperature of the coolant from chiller in °C

9.1.1 Specific Power Required

After calculating the tonnage, the specific power consumption is calculated. It is the ratio of the total power required and the tonnage required for cooling the area.

Specific power consumption = KW/TR

9.1.2 Coefficient of Performance

Coefficient of performance is also a measure for finding out the efficiency of air conditioning system. Coefficient of performance can be calculated by the following formula:

COP = Te/(Tc – Te)

Te = Temperature of evaporator

Tc = Temperature of condenser

Another measure of coefficient of performance is the ratio of rate of heat removal and input power to compressor in KW.

9.2 Conclusion

Refrigeration system is an area where scope of energy conservation is vast. For example, just by changing the temperature setting by one degree, the switch on and off time of compressor can be changed and in this way saving in energy takes place. Just a few salient points which are related with the performance of the refrigeration system are touched in this chapter.

9.3 Software

The software is capable to perform calculations related with energy saving in refrigeration system.

9.3.1 Objective

The objective is as mentioned below:

- To build a database of the air conditioning system for which energy auditing is required.
- To generate query for retrieval of the data from the database for analysis of system.
- To transfer the retrieved information on the analysis screen and to perform calculations related with energy saving.
- To take a printout of the results of calculations.

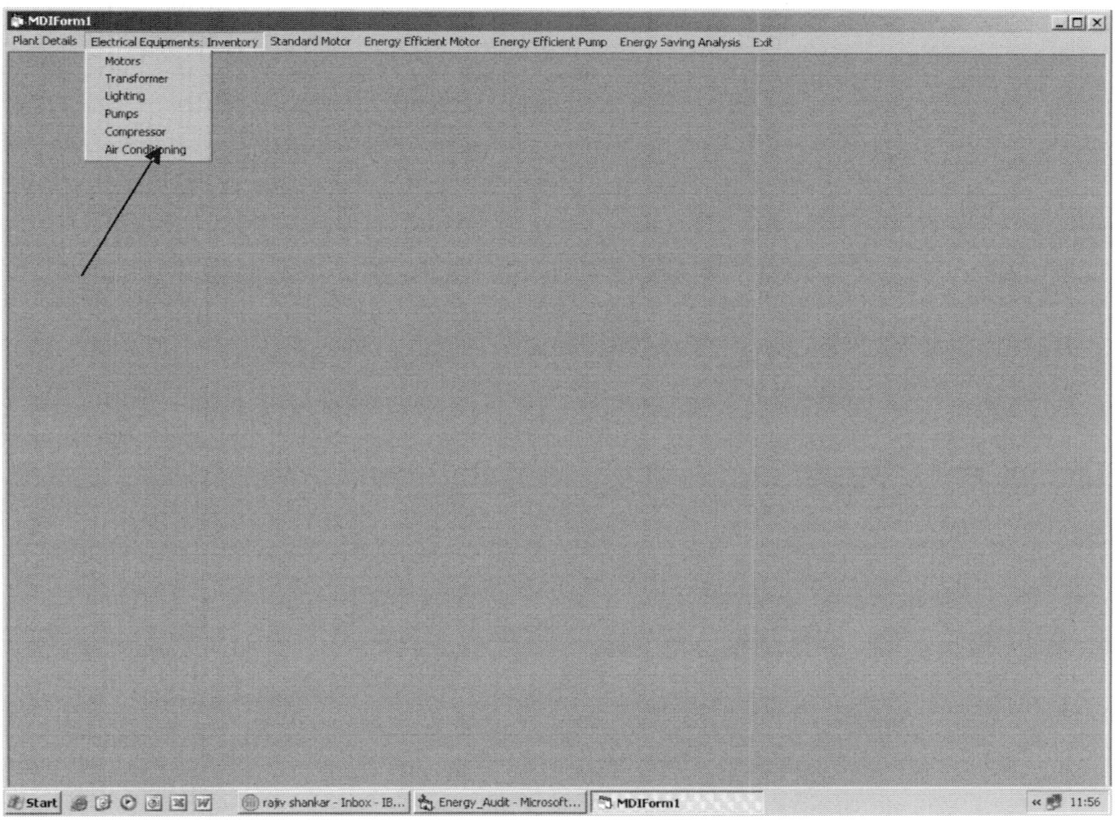

Fig. 9.0 Main screen

9.3.2 Procedure

- Click on "Electrical Equipments: Inventory" and after that click on "Air Conditioning". This will result in opening of a screen as shown in Fig. 9.1. Fill in the required information in the respective boxes.

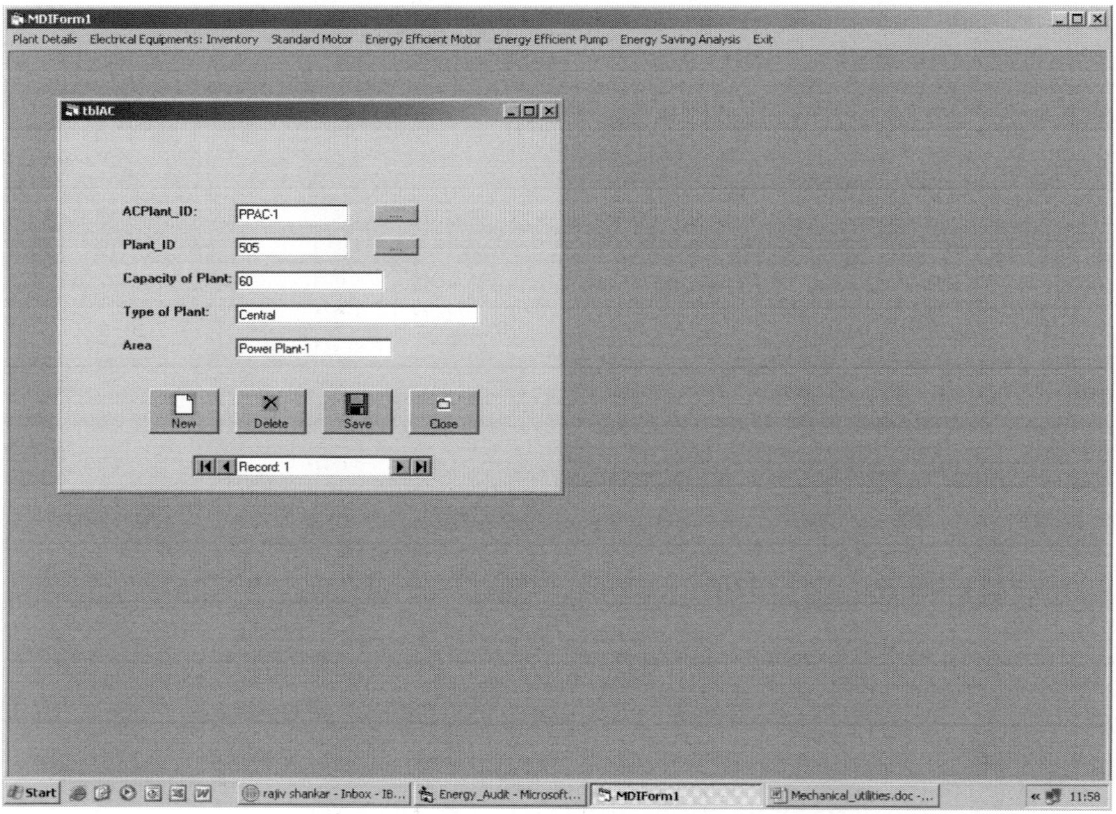

Fig. 9.1 Preparation of data bank

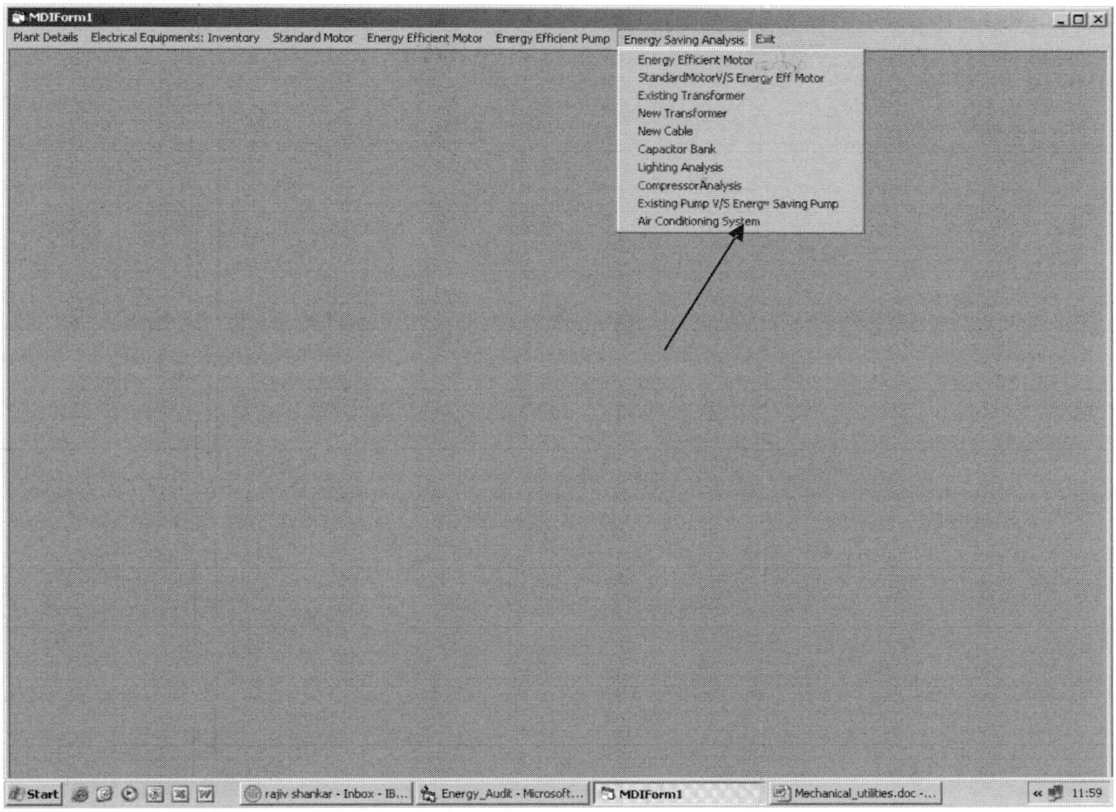

Fig. 9.2 Main screen with submenu for analysis of ACplant

- Click on "Energy Saving Analysis" button a submenu will appear. Click on "Air Conditioning System" of the submenu as shown in Fig. 9.2 by an arrow. A screen as shown in Fig. 9.3 will appear.
- Click on "Database" button as shown by an arrow in Fig. 9.3.

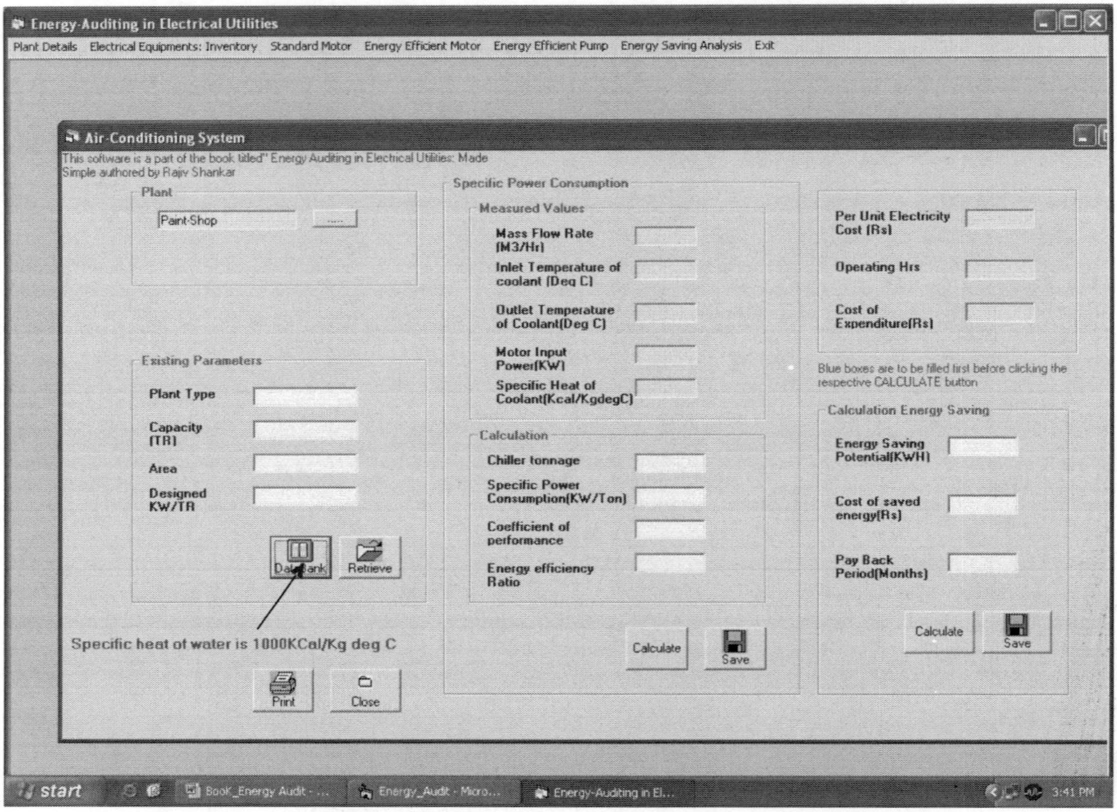

Fig. 9.3 Analysis of air conditioning system

A query screen will open up. The query is generated based on the plant and the capacity of the air conditioning system, as shown in Fig. 9.4. To select an air conditioning system click on "ACPlant-ID". This clicking action will transfer all the parameters to the main screen for analysis.

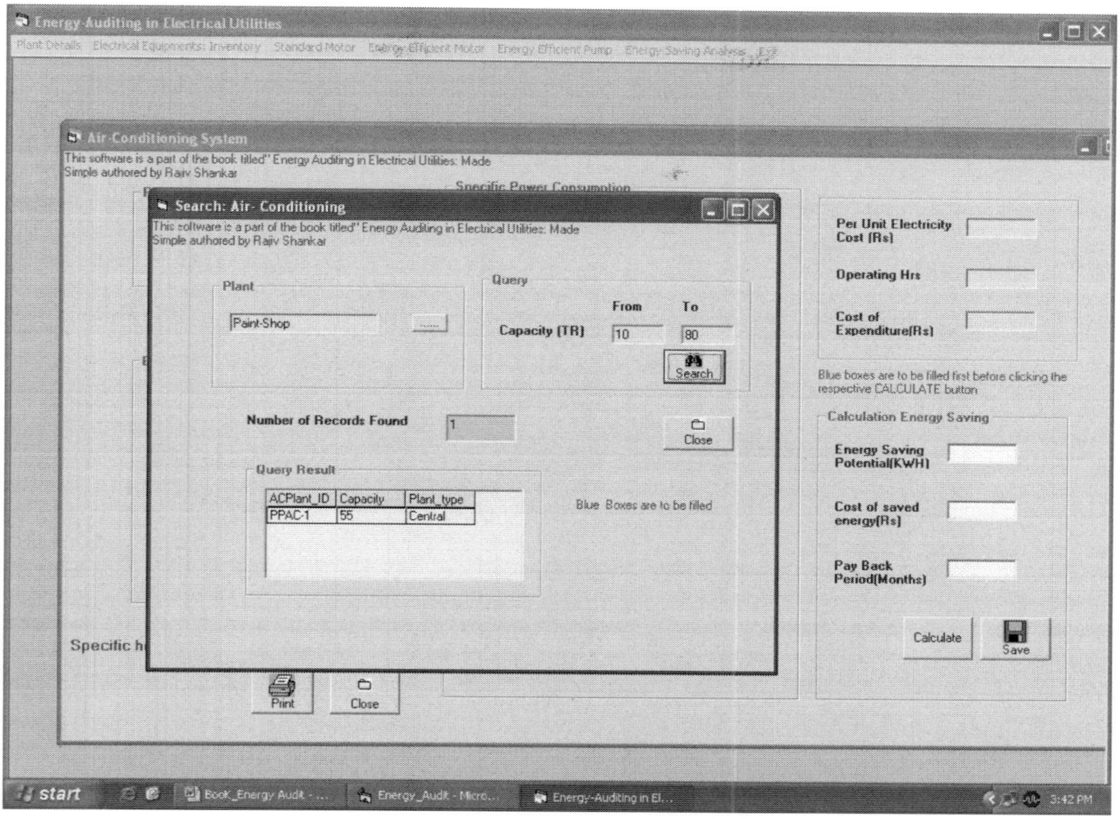

Fig. 9.4 Retrieving of air conditioning data

- Now, click on "Retrieve" button as shown in Fig. 9.5 by an arrow to retrieve the information.

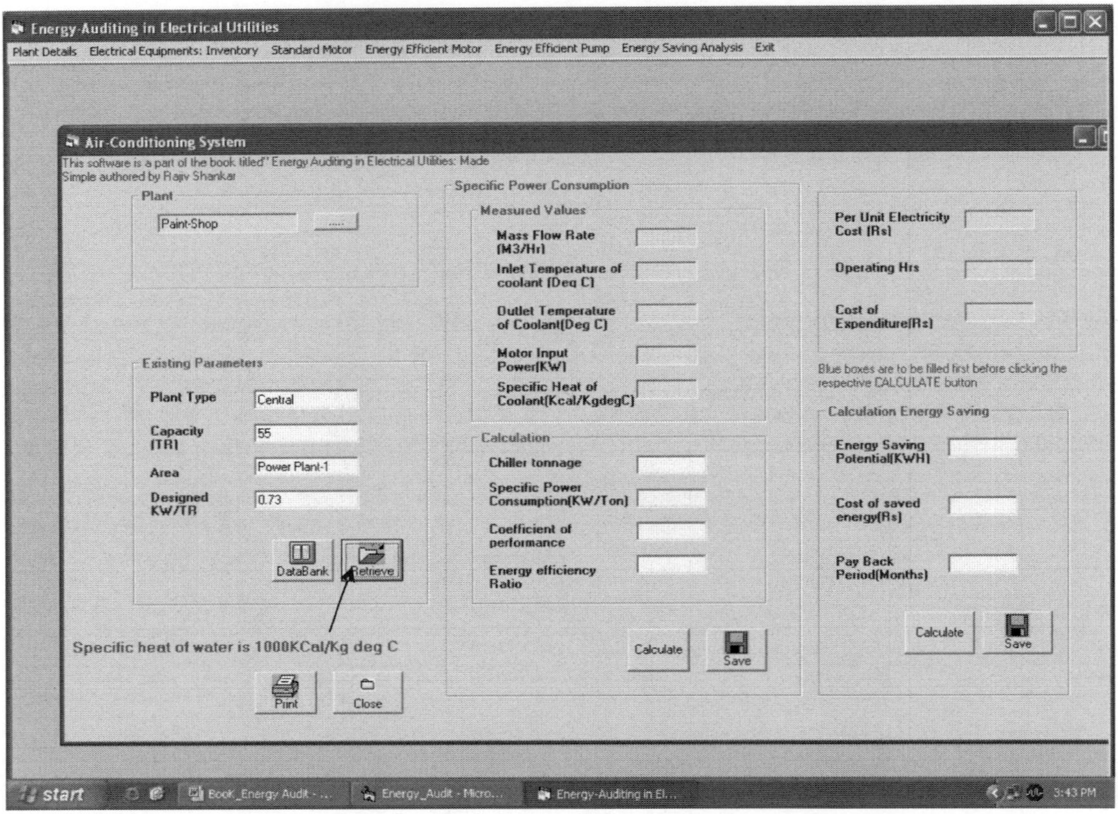

Fig. 9.5 Transferring data to air conditioning analysis screen

- Fill in the measured parameters in the respective boxes of frame titled as "Measured Values" as shown in Fig. 9.6. Click on "Calculate" button to calculate the chiller tonnage, specific power consumption, coefficient of performance and energy efficiency ratio. These are all important parameters and required to be calculated for evaluation of performance of any air conditioning system. The same screen also has facility for calculation of energy saving potential by comparing it with the designed value of specific power consumption.
- Fill in the cost of electricity, running hours of air conditioning system and the cost of equipment installed or any modification done to improve the performance of the air conditioning system. Click on "Calculate" button to calculate the energy saved and the payback period also.
- Click on "Save" button to save the results and then click on "Print" button to take the print of the results.

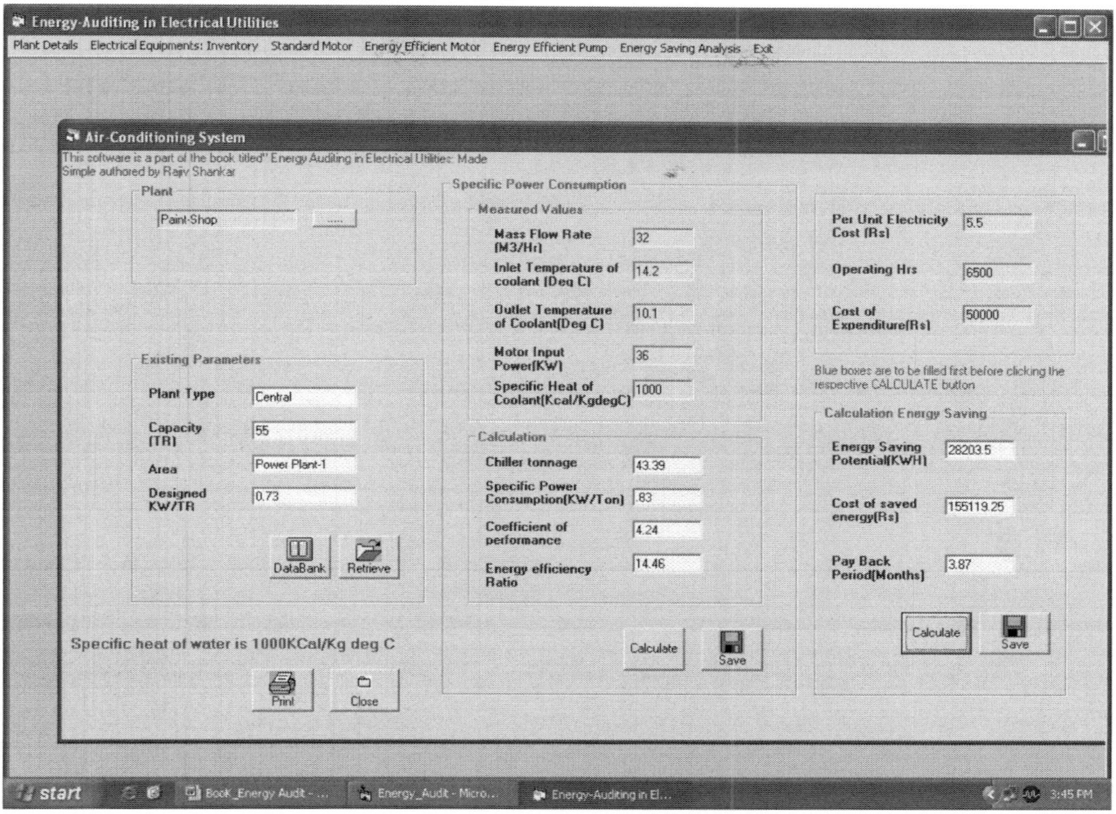

Fig. 9.6 Energy saving analysis

Installation of Software

1. The CD-ROM contains a folder "Energy-Audit"
2. Copy the "Energy-Audit" folder from the CD to C: drive
3. Open the "Energy-Audit" folder
4. Open the "Package" folder and double click on "Setup.exe" icon.

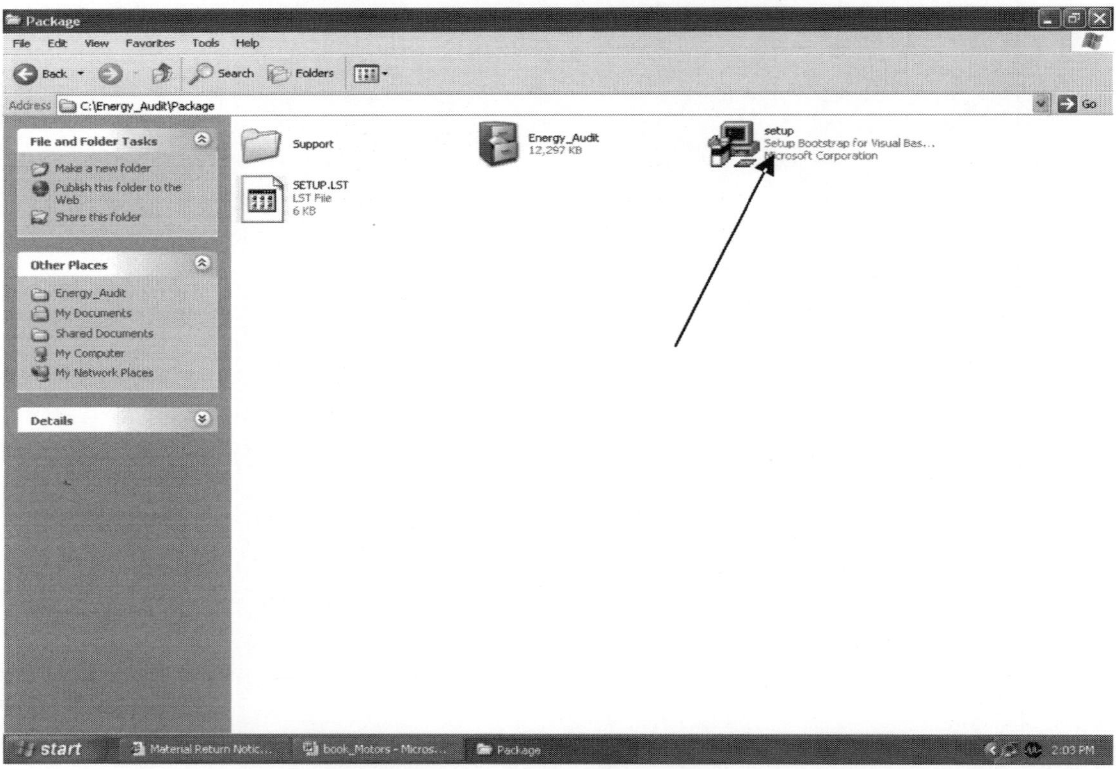

5. Energy-Audit Setup screen will open up. Click on "OK" button.

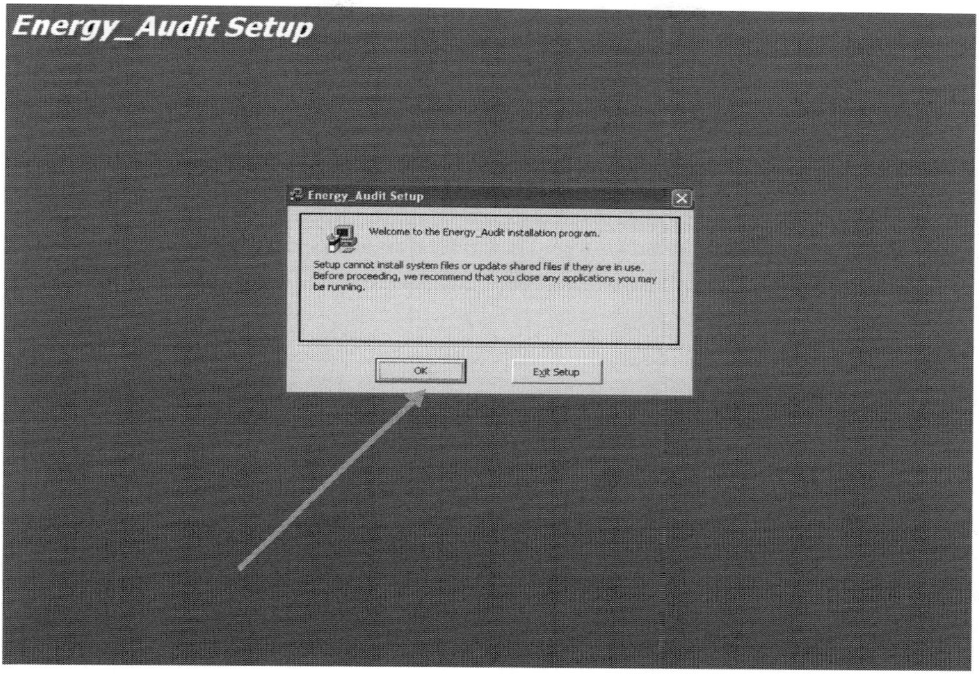

6. Click on "Installation" button.

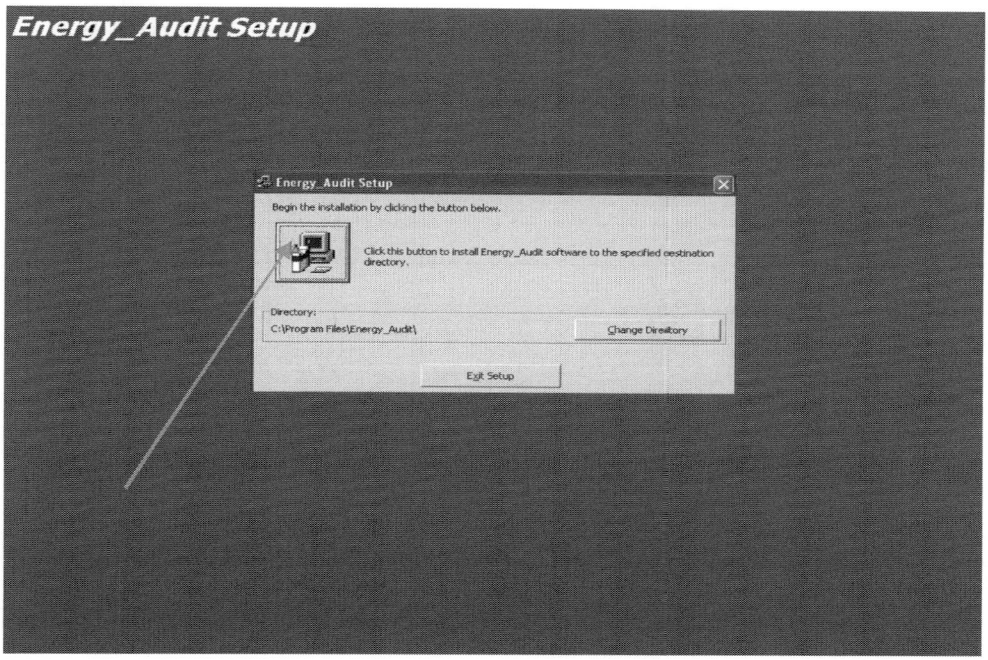

7. Click on "Continue" button of setup screen.

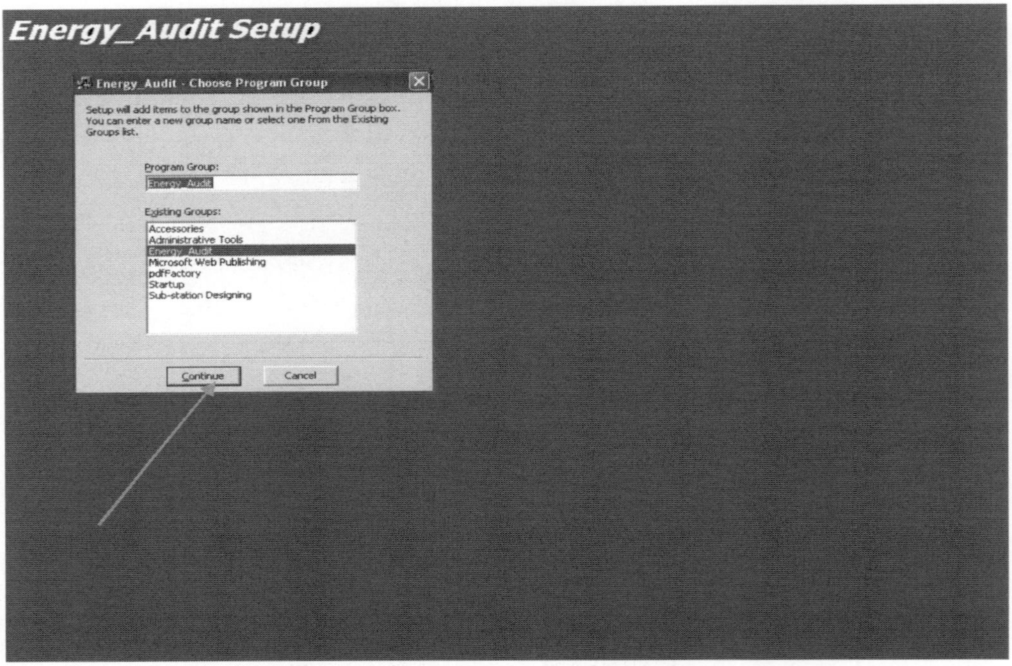

8. Program files will be copied in the desired location.

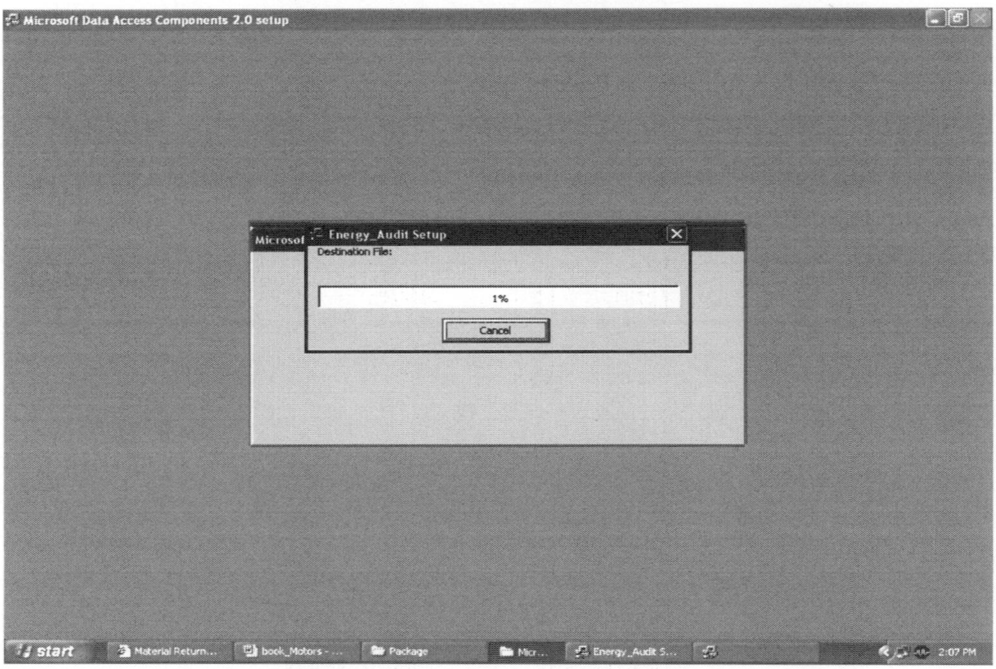

9. Message will appear "Energy-Audit Setup" was completed successfully. Click on "OK" button.

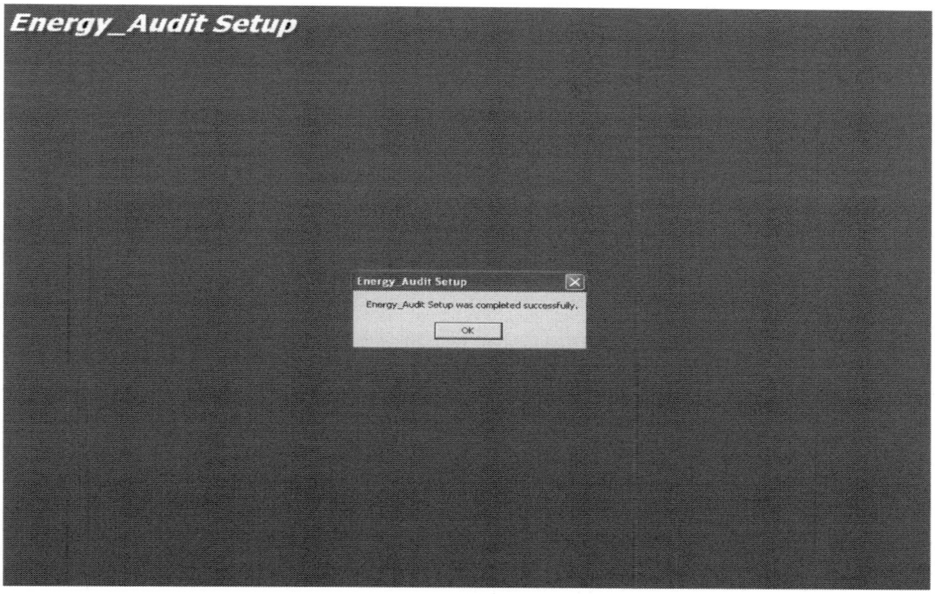

10. Go to program and click on "Energy-Audit" to run the program.

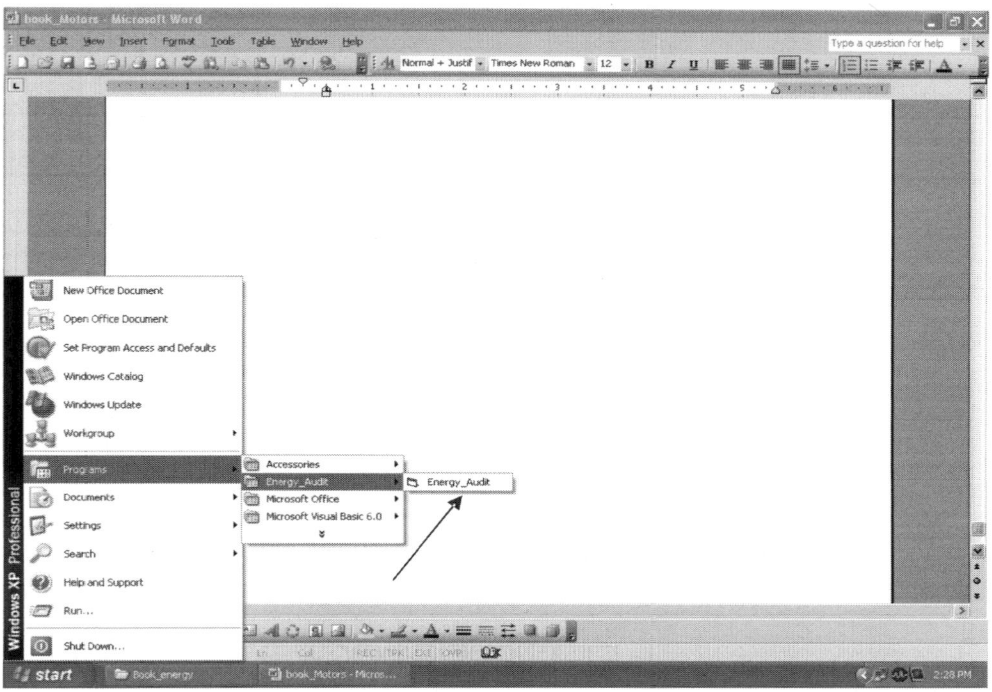

Index